INDIGENOUS PEOPLE AND POLITICS

Edited by
David Wilkins
University of Minnesota
Franke Wilmer
Montana State University

A ROUTLEDGE SERIES

Inventing Indigenous Knowledge
Archaeology, Rural development, and the Raised Field Rehabilitation Project in Bolivia

Lynn Swartley

Routledge
Taylor & Francis Group

NEW YORK AND LONDON

Published in 2002 by
Routledge
711 Third Avenue
New York, NY 10017, USA

Published in Great Britain by
Routledge
2 Park Square,
Milton Park, Abingdon
Oxfordshire OX14 4RN

First issued in paperback 2016

Routledge is an imprint of the Taylor and Francis Group, an informa business

Copyright © 2002 by Taylor & Francis Books, Inc.

Library of Congress Cataloging-in-Publication Data

Swartley, Lynn.
  Inventing indigenous knowledge : archaeology, rural development, and the raised field rehabilitation project in Bolivia / Lynn Swartley.
    p. cm. — (Indigenous people and politics)
    ISBN 0-415-93564-4
    1. Indians of South America—Agriculture—Bolivia. 2. Aymara Indians—Agriculture. 3. Rural development projects—Bolivia. 4. Terracing—Bolivia. I. Title. II. Series.
  F3320.1.A47 S93 2002
  631.4'55'098412—dc21   2002031704

ISBN 13: 978-1-138-97331-2 (pbk)
ISBN 13: 978-0-415-93564-7 (hbk)

Dedicated to my parents, Howard and Janet Swartley

Contents

Acknowledgments

THIS WORK WAS MADE POSSIBLE BY THE GUIDANCE, SUPPORT, AND FAITH OF numerous family, friends, colleagues, mentors, and teachers. First and foremost, I would like to thank the people of Wankollo who let me work in their community, patiently answered my questions, and took the time to explain to me their goals, aspirations, and desires. I give my sincere thanks to all of them, but most especially to my assistant, Leonardo Laura, and my *compadres*, Alejandro Choque and Victoria Laura de Choque. Second, I would like to thank the personnel at the *Fundación Wiñaymarka* who assisted me by providing interviews and unpublished reports. Third, I would like to thank *Ayni Tambo,* who worked with me during June and July of 1994, and when I returned in 1996. I would also like to thank the National Institute of Archaeology and Anthropology in Bolivia for assisting me during fieldwork. I want to thank all of the archaeologists, who helped me, talked to me, and shared their work with me, but particularly John Janusek, who let me tag along on numerous occasions and projects. I would also like to thank Joanne Harrison and Deborah Blom for their friendship in the field. Finally, I would like to express my warmest gratitude to my friend, Claudia Heckl, who provided shelter, support, friendship, and laughter while I lived in Bolivia.

At the University of Pittsburgh, I give my deepest gratitude to the four members of my dissertation committee. Harry Sanabria, my advisor, was a mentor in every sense of the word, guiding me from the moment I stepped onto campus, and spending countless hours reading numerous drafts of grant proposals, outlines, and my dissertation. His thoughtful and thorough comments have made me a better scholar. Kathleen DeWalt encouraged me and supported my efforts without fail. I am grateful to have her as a role model. Marc Bermann introduced me to the raised fields project in Bolivia and has supported my research, while offering insightful critiques of my work. Bill DeWalt made time in his very busy schedule to offer his encouragement, advice, and support.

I would like to thank my friends and colleagues at the University of Pittsburgh: Maggie Smith Brackett, the best of friends, who wrote sustaining letters from India while I was in the field, and Carolyn Myers, who also wrote to me from afar (the Solomon Islands) and was my comrade since entering the department together. I would also like to thank the support and friendship of Michelle Madsen-Camacho, who was a constant source of inspiration during writing. Finally, my dear friend Dawn Barrie, who has seen me through many years in Pittsburgh and endured the 6 hour drive from Ithaca on so many occasions to offer a shoulder to cry on, and the love and laughter of true friendship.

Friends and colleagues at Rutgers University provided a stimulating and supportive environment where I was able to finish this manuscript. I give them my gratitude for their support during the difficult first year of my post-doctoral fellowship. I would particularly like to thank the directors of the 2001–2003 Project on "Industrial Environments" at the Center for Historical Analysis, Susan Schrepfer and Philip Scranton.

I need to acknowledge the financial support of several institutions. This research was supported by grants from the National Science Foundation, the Fulbright Institute of International Education, and the Inter-American Foundation. I would like to thank the Center for Latin American Studies at the University of Pittsburgh for providing language training with a summer FLAS and a summer research grant. Also, I wish to thank the Lithuanian Room at the University of Pittsburgh for an additional summer travel grant.

Finally, I wish to thank my family, especially my mother Janet Swartley and my father Howard Swartley, for believing in me and always encouraging me to follow my dreams and to pursue all of life's adventures. Lastly, I give my gratitude to my late husband Ted Ahlen (1968–2001), who supported me with love and patience during the writing of this book.

INVENTING INDIGENOUS KNOWLEDGE

.

CHAPTER ONE

Introduction

RAISED FIELDS ARE A FORM OF INTENSIVE AGRICULTURE THAT WAS IN widespread use by the inhabitants of the Lake Titicaca Basin during pre-Hispanic eras. However, it was not until the advent of high resolution aerial photography, which provided detailed images of the wide distribution of these pre-Hispanic raised fields, that research on this ancient technology began to generate increasing interest among geographers and archaeologists (Smith et al. 1968; Erickson 1988a). In the early 1980s, North American archaeologists working in the Lake Titicaca Basin of Peru and Bolivia led teams in the excavation of pre-Hispanic raised fields (Erickson 1985, 1987, 1988a; Kolata 1986, 1991, 1993). Raised fields are a system of raised bed agriculture that was practiced extensively throughout the Western hemisphere in the pre-Hispanic past (Turner and Denevan 1985; Smith et al 1968). Remnants of pre-Hispanic raised fields are found throughout North and South America, though they are most common in Central and South America (Parsons and Bowen 1967; Turner 1974). Raised fields had virtually disappeared by the arrival of the Spaniards (Graffam 1990, 1992; Seddon 1994) with the exception of the floating gardens of the Aztecs in Mexico (Armillas 1971; Coe 1964).[1]

Raised fields are elevated platform beds that raise plants above the waterline in the seasonally inundated land of the high plains that surround Lake Titicaca in Peru and Bolivia (see Figure 1—Wankollo Raised Fields). A key component in the construction of the fields is the building of canals that surround the fields on all sides. The canals collect and hold excess water, retain sediments and soil eroding from the planting surface, and generate nutrients for maintaining soil fertility on the fields by creating a green manure in their stagnant waters (Erickson 1988a; Carney et al. 1996). Researchers have also found that if adequate water levels are maintained in the canals, the water retains solar heat during the day that is released at night creating a microclimate effect that helps protect plants from frost damage on the excessively frost prone plains of the lake basin (Kolata and Ortloff 1996; Sánchez de Lozada 1996).

By the mid-1980s, excavations of raised fields in the Lake Titicaca Basin led to practical experiments reconstructing the fields, with the goal of better understanding the function and productivity of the fields in order to identify the social organization and economic potential of pre-Hispanic societies. These initial experiments produced amazingly positive results, with the newly reconstructed raised fields offering extended protection to young plants from frost and producing impressive initial yields, often several times as much as the regular dry flatland fields that contemporary farmers cultivated (Erickson 1988a; Kolata et al. 1996). The high production yields of these early experiments engaged the interest of agricultural researchers and social scientists.

However, there were other factors that gave raised fields an instant appeal to researchers and development workers, both in Bolivia and abroad. Not only were the fields highly productive per unit of land cultivated, but they were also considered an "indigenous" and ancestral form of knowledge that had been rediscovered through the work of geographers and archaeologists. Since the fields were labeled as "indigenous agriculture," they came to be considered an aboriginal form of agriculture that was native to the lake basin. This stands in sharp contrast to the "green revolution" agriculture that was perceived as a foreign, Western style of agriculture. Examples of Western agriculture include farming machinery and chemical soil additives, as well as Western methods of agriculture such as monocropping. The rediscovered raised field system was promoted as indigenous, not needing any chemical soil additives, and using only manual labor. The ancestral and indigenous pedigree conferred to the raised fields led researchers to uncritically assume that the fields were ecologically compatible with the lake basin environment; the fields were native to the lake basin, thus it was assumed that they were a natural fit with the social and physical environment of the contemporary agriculturalists.

The representation of the fields as indigenous agriculture also coincided with the "indigenous knowledge" and "sustainable development" trends in academic theories of development. The concept of sustainable development attempts to incorporate environmental concerns into the development agenda by making them integral facets of development, development policies, and development projects. This new environmental component of development policy came to be called sustainable development. Linked to this trend towards sustainable development was a coinciding interest in theories of development that privileged local or "indigenous" culture and knowledge. As I will argue, the linking of sustainable development and theories that privilege indigenous knowledge is partially due to North American and European preconceptions of Native Americans as being innate environmentalists and conservationists. Krech (1999) argues that North Americans constructed an image of Native Americans as "ecological Indians" who understood their environment better than Europeans, and whose cultural practices were more closely tied to nature conservation.

As North American and European development agencies were celebrating indigenous knowledge as a possible solution for the new sustainable development trend, in Latin America there were increasing demands for indigenous social rights. Indigenous social movements were widespread across Latin America in the 1980s (Hale 1994), with the *Katarista* movement led by Aymara speaking intellectuals gaining political power in Bolivia in the late 1980s (Albó 1987, 1994). As the *Kataristas* gained power in Bolivia, many of the traditional parties of the country soon began adopting and integrating indigenous symbols and discourse into their own political platforms (Albó 1994; Rivera Cusicanqui 1993; Van Cott 2000a).

These simultaneous and interconnected trends in environmentalism, sustainable development, and interest in indigenous knowledge in Europe and North America, as well as the concomitant rise in indigenous social movements in Latin America, created an ideal context for raised field agriculture to capture the attention of development workers and development agencies in Bolivia, the U.S., and Europe. By the late 1980s, several archaeologists and development NGOs, both in Peru and in Bolivia, were interested in reconstructing and rehabilitating raised fields in contemporary Lake Titicaca Basin communities. In Bolivia, the NGO *Fundación Wiñaymarka*, headed by the director of the Bolivian National Institute of Archaeology, led efforts to rebuild the raised fields. By the early 1990s, this NGO had recruited 55 Lake Titicaca Basin communities in Bolivia to participate in the raised field project and had reconstructed over 94 hectares of raised fields.

The reconstructed raised fields were hailed as an instant success story and news of the wondrous rehabilitation of ancient raised fields made it into North American newspapers, a National Geographic special, and into anthropology and archaeology text books for college students in the United States. In all of these public forums, the fields were represented as offering bountiful crop production far superior to the techniques of contemporary farmers of the Lake Titicaca Basin. A number of scholarly works (Erickson 1992a; Kolata et al. 1996) hypothesized that this ancient indigenous technology contained a possible solution for increasing the agricultural production of contemporary smallholder farmers in the Lake Titicaca Basin.

Raised fields were portrayed as superior to the production increases achieved by the costly application of green revolution chemical fertilizers and pesticides. The high productivity on raised fields was attained through the application of an "all natural" and "organic" method of cultivation that had been practiced by indigenous peoples who had inhabited the Lake Titicaca Basin over a millennium ago. Raised fields cost little to build (except for the manual labor of the farmers), needed few production inputs besides seed and access to land, and supposedly were capable of continuous cropping without the need for long fallow periods as normal potato fields required.

In addition to the economic appeal of the raised fields, they also acquired symbolic value as a symbol of indigenous culture and heritage. In Bolivia, the raised fields were associated with the pre-Hispanic Tiwanaku civilization of the southern Lake Titicaca Basin. The epicenter of Tiwanaku, and what has been called the "Tiwanaku heartland" (Kolata 1986), lies within the borders of the current nation-state of Bolivia. The site and civilization of Tiwanaku are a symbol of the contemporary Bolivian nation-state. The pre-Hispanic raised fields, which were popularly associated with the Tiwanaku civilization, also became symbolic of national cultural heritage. Furthermore, the fields were considered in Bolivia to be an agricultural method of *"los descendientes de los antiguos tiwanakotas"* (the descendents of the ancient people of Tiwanaku)(Rojas-Velarde 1996:1), because most archaeologists believe that the Aymara people who now live in the Lake Titicaca Basin are the descendants and rightful inheritors of the former Tiwanaku polity (Browman 1994).

I argue that the raised fields were an "invented tradition" created by archaeologists who were investigating and experimenting with the ancient remains of raised fields. I use the term invented tradition to emphasize the set of practices and social organization that archaeologists proposed for building and cultivating the raised fields in contemporary communities. The social organization and methods of rehabilitating raised fields reflected the values of the archaeologists, while overtly implying that these values shared continuity with the pre-Hispanic past (Hobsbawm 1983). I argue that the contemporary practice and social organization of raised field development reflected preconceived notions of indigenous peoples and the peasantry in Latin America. Raised field agriculture was not an actual system of knowledge that was intact and in practice by contemporary inhabitants of the Lake Titicaca Basin. The contemporary practice of raised field agriculture was an invented tradition, which maintained ethnic and class boundaries through the symbolic appropriation of the past.

In addition to the symbolic value of raised fields, there were also social and economic factors that caused problems for the contemporary rehabilitation of raised field agriculture. When proposing that raised fields be rebuilt and rehabilitated, archaeologists placed much more emphasis on the notion that raised fields had been used extensively over a long period of time in the distant past. Thus, archaeologists concluded that raised fields were ecologically well adapted to the harsh, high altitude environment of the Lake Titicaca Basin. Less attention was given to the social, political, and economic contexts of raised field use, which had changed drastically since the height of pre-Hispanic raised field cultivation in the past. Therefore, a second goal is to explore and explain some of the social, political, and economic factors that conflict with and constrain raised field cultivation in contemporary Bolivian communities.

I arrived in Bolivia in May of 1994 to do initial reconnaissance research for my doctoral dissertation on the social economics of raised field cultivation as it was being practiced in contemporary Bolivian communities. Once in Bolivia, I joined a team of researchers hired by the Inter-American Foundation (IAF), which had been one of many organizations that had funded various raised field rehabilitation projects in Peru and Bolivia. The research team was contracted by IAF to do a survey of a dozen communities that had participated in the rehabilitation project conducted by the NGO *Fundación Wiñaymarka,* to assess the implementation and relative success of the rehabilitation project. During the course of fieldwork with the team, which was carried out over my two-month stay in Bolivia, it became apparent that the raised fields had not been quite the long-term success that initial experiments and reports had indicated. Many of the communities that had built raised fields in the late 1980s and early 1990s had seen sharp declines in production in recent years, and some raised fields had already been abandoned and returned to fallow.

By March of 1996 when I returned to Bolivia to begin dissertation fieldwork, the NGO *Fundación Wiñaymarka,* which had had an office and several workers in 1994, was no longer building raised fields and had completely disbanded. Upon my return to the countryside and the Lake Titicaca Basin, it was immediately apparent that all of the communities that had participated in the raised field rehabilitation project with the NGO *Fundación Wiñaymarka* had discontinued cultivation of the raised fields. Academic and development interest in the raised fields also seemed to have tapered off. For example, in 1994 Bolivian archaeologists and rural development workers who knew of the raised fields had been eager to talk about them and had discussed them with some pride as a Bolivian agricultural success story. However, by 1996 their enthusiasm had waned. In 1996, I spoke to many of the same Bolivian archaeologists and development workers with whom I had spoken to previously and who had been interested in the raised fields. Yet many of those who I re-interviewed in 1996 were no longer interested in the raised field project and in recovering an ancient indigenous technology, even when that knowledge was depicted as a uniquely Bolivian heritage.

Why had the raised fields been abandoned? Why did the development groups who had implemented the raised fields in Bolivia consider them a failure in 1996 when they had produced such prodigious initial yields? Why was the development of raised fields as a form of indigenous knowledge so appealing in the late 1980s and early 1990s, but less so by 1996? It is the asking and the answering of these questions that forms the basis of this work.

This book is not an ethnography of "the Aymara" nor is it strictly an account of rural economy or rural life in Bolivia. It is a multi-sited and multivocalic investigation of the dynamic social, political, and economic processes in the creation, implementation, and eventual demise of an agricultural development project. The raised field rehabilitation project attempted to introduce a pre-Hispanic agricultural method into contemporary Lake Titicaca Basin communities in Bolivia. Rather than simply investigating why the project did not work and trying to put together the broken pieces of the puzzle from the perspective of a failed development project, I take a broader look at the project from the perspective of Bolivian political and economic history. I establish the contexts of the raised field project by exploring the social, political, and economic trends leading up to the implementation of the project, such as the rise of ethnic politics in Bolivia, the economic crisis and structural adjustments of the mid-1980s, and the concurrent shift towards international development agendas emphasizing the environment and ecological sustainability.

To find a complete and detailed explanation of the rehabilitation project, one must unravel the story of how and why the raised field rehabilitation project came to be considered potentially "successful" and "appropriate technology" for the Bolivian highlands in the first place. One has to ask the question, why were raised fields so appealing to North America researchers and Bolivian NGOs during the 1980s and early 1990s? What was the mystique and appeal of resurrecting an indigenous knowledge that had held such a captive audience with the archaeologists who experimented with the fields and the development workers who attempted to reconstruct them on a large scale?

This book explores more than just the material factors in the life span of a development project. It considers the symbolic and cultural aspects of the project through an analysis of how academics and development groups represented raised fields and indigenous peoples. By interpreting the messages embedded in the representation of the raised field rehabilitation project, I reveal how these messages conflicted with the ambitions, goals, and desires of the Aymara smallholder agriculturalists of the Bolivian Lake Titicaca Basin. Using a diachronic approach, I examine the life cycle of the raised field rehabilitation project in Bolivia. From the "rediscovery" of this so-called lost technology, to its climax as a model of sustainable development and applied archaeology, to the forgotten and abandoned fields that dot the high plains today, raised fields have come and gone from Bolivia once again.

This work is structured by two fundamental questions. First, why did the raised field development project garner so much academic interest, media attention, and development support? The answer lies in unraveling the historical formation of several convergent processes such as the social and economic contexts of agriculture in Bolivia, emergent indigenous political movements that drew on ethnic claims to the past, and international development trends that culminated in the sustainable development of the late 1980s. Chapter Four analyzes the representations of the raised fields by development groups and academics that portrayed the fields as "sustainable agriculture," "indigenous knowledge," and "appropriate technology."

The second question that frames this book is, how did the representation of raised fields—for example as "sustainable agriculture" and "indigenous knowledge"—conflict with the rural economics of agriculture in the Bolivian Lake Titicaca Basin? This question draws on the representations produced by academic researchers and development groups, and compares them with the economics of agriculture in the Bolivian Lake Titicaca Basin. Drawing on ethnographic fieldwork in a community that participated in the raised field rehabilitation project sponsored by the NGO *Fundación Wiñaymarka*, this work examines agricultural production, land tenure, and access to labor for agriculture. Ultimately, I offer an explanation for why raised field agriculture was abandoned in communities that participated in the Bolivian rehabilitation development project after only 3 to 4 years of cultivation.

REPRESENTING RAISED FIELDS AS INDIGENOUS KNOWLEDGE

I demonstrate how archaeologists and development workers represented raised fields as an indigenous knowledge—a form of knowledge defined in contrast to and distinct from Western scientific knowledge. Yet as many philosophers and social scientists have argued, the social foundations of knowledge and knowledge systems are not something that researchers should take for granted, but rather they are conceptual constructs that should be explained (Apffel-Marglin 1996; Hacking 1999; Meja and Stehr 1990; Worsley 1997). It is problematic that researchers and development workers labeled and represented the raised fields as an indigenous knowledge without exploring how this label and its social meanings are influenced and manipulated by the politics of development and indigenous social movements.

One long-standing philosophical debate pits Western scientific knowledge, currently the dominant form of knowledge, against other non-Western, local, or indigenous systems of knowledge (Apffel-Marglin 1996). Social constructionist works that explore the Western concept of knowledge privilege the role of culture and power in describing how a system of knowledge is defined and how its boundaries are maintained vis-à-vis other knowledge systems. For example, McCarthy writes that "*knowledge* is best conceived and studied *as culture*, and the various types of social knowledge communicate and signal social meanings" (italics in original)(1996:1). For McCarthy, this field of study is best described as the "sociology of knowledge and culture," since he makes a direct connection to the cultural basis of all systems of knowledge as they invoke culturally shared symbols, meanings, and practices. As Hacking (1999) points out, such social constructionist approaches to the sociology of knowledge are "liberating" in that they reveal that social representations and social labels defined by the dominant discourse of knowledge are not fixed and inevitable categories.

Connected to this research on the sociology of knowledge is the increasing interest in indigenous knowledge—sometimes called indigenous technical knowledge or local knowledge—which has garnered much attention with development institutions and social scientists, particularly anthropologists working with international funding institutions and NGOs. Indigenous knowledge was championed in the development sector by the work of Michael Warren (1989a, 1989b, 1991, 1999). Prior to his work, indigenous cultural practices were generally disregarded in development under the assumption that they were backward and inferior to Western economic development. Warren's (1991) discussion paper for a seminar series on the sociology of natural resource management at the

World Bank, introduced the concept of indigenous knowledge into the mainstream of development theory. In this work, Warren defined indigenous knowledge as local knowledge that is "unique to a given culture or society" and that forms "the basis for local-level decision-making in agriculture, health care, food preparation, education, natural-resource management, and a host of other activities in rural communities" (1991:1). More generally, indigenous knowledge is defined in opposition to Western scientific knowledge, thus it has the similar polemic dichotomy between dominant Western knowledge and other cultural forms of knowledge.

Interest in indigenous knowledge systems is directly linked to the shift in theories of economic development towards ecological and economical sustainability. The concept of "sustainable development" emphasizes models of economic development that foster economic growth without jeopardizing the ecological foundations of natural resources and future economic growth. Sustainable development has its roots in the conservation and environmental movements of the 1960s and 1970s (Adams 1990). Sustainable development came into the mainstream of development discourses with the release of the Brundtland report in 1987, which retrieved the idea sustainable development from the fringes of development theories and featured it as the United Nations' new mandate for economic development. The Brundtland report defined sustainable development as "development that meets the needs of the present without compromising the ability of future generations to meet their own needs" (Brundtland 1987:43).

Critics of the indigenous knowledge approach to sustainable development have pointed out that all knowledge is "situated knowledge" and that indigenous knowledge is particularly context specific (Nazarea 1999). By describing indigenous knowledge as situated knowledge, the implication is that culture and knowledge cannot be separated. Indigenous knowledge, therefore, often has little use outside of its immediate cultural contexts since it usually cannot be replicated.

On the other hand, critics of Western scientific knowledge have emphasized the ways in which Western science strengthens and supports the interests of powerful elites and elite classes (DeWalt 1999; Foucault 1972). They argue that no single form of knowledge can be defined as a universal knowledge system for classifying the world. Instead, critics argue that there are many forms of knowledge, among them indigenous knowledge, which are useful for categorizing and understanding the world (Nazarea 1999). This underlines the role of power in the contestation over who gets to count their form of knowledge as the dominant or prestige form of knowledge in a given contexts (Nazarea 1999). Therefore, what is at stake is the control and dissemination of knowledge, particularly since knowledge in the modern world-system has become a commodity (Brush 1993).

Some supporters of indigenous knowledge systems are concerned that indigenous knowledge is being lost, with the implication that this is much to the detriment of local indigenous peoples (Hunn 1999; Soleri and Smith 1999). Hunn describes the situation as dire writing that "bodies of traditional knowledge are gravely threatened, in imminent danger of going to the grave with the present generation of elders" (1999:23). Thus it is argued that indigenous knowledge should be preserved (Hunn 1999) and safeguarded (Posey 1999) against the encroachment of Western knowledge and culture. Others have advocated that indigenous knowledge should be protected through the granting of intellectual property rights (Brush 1993; Posey 1999; Stephenson 1999) and that this indigenous knowledge should be compensated when it is used by Western science, particularly in the case of drug discoveries and medicines (Moran 1999).

More recently, some social scientists have approached the subject of indigenous knowledge systems with a more even-handed critique (Agrawal 1995; DeWalt 1994, 1999). For example, DeWalt (1994, 1999) discusses the strengths and weaknesses of both Western scientific knowledge and indigenous knowledge systems. He criticizes those in the development field who have taken on an almost "missionary fervor" about the virtues of indigenous knowledge systems. Instead, DeWalt proposes a framework for integrating scientific and indigenous knowledge systems to improve agriculture and natural resource management. DeWalt's framework is a practical approach they may help Western trained scientists evaluate and integrate indigenous knowledge into their Western scientific models of development.

According to the framework proposed by DeWalt, development projects should integrate both indigenous and scientific knowledge without relying solely on the findings of one or the other. He also notes that both systems of knowledge are influenced and controlled by the way people think and understand their social world. However, DeWalt (1999) acknowledges that there is "an implicit assumption" that science is "value free" and "socially and culturally neutral." Therefore, all knowledge is constructed through a dialectical process between the author and the subject, or the observer and the observed. Any research derived from this knowledge is part of a political process (Agrawal 1995; DeWalt 1999).

DeWalt's (1999) framework is useful in order to educate Western trained development workers in the understanding and utilizing of non-Western cultural practices and knowledge. By labeling cultural knowledge and other aspects of culture as indigenous knowledge, supporters of the indigenous knowledge approach in rural development are attempting to give local indigenous cultures a new and different image, one that is more easily understood and palatable to Western trained development workers. The development profession has always had an eye for catch phrases and slogans, and the concept of indigenous knowledge artfully categorizes all cultural differences previously viewed in the development sector as "traditional" and "backward" under a new slogan. Thus the new indigenous knowledge slogan has a more 21st century multicultural appeal to environmentally conscious Westerners.

Therefore, the concept of indigenous knowledge still emphasizes the differences between Western and non-Western knowledge, but without the stigma of being traditional and backward—two terms that have very negative connotations in the development profession. However, DeWalt's (1999) framework continues this dualistic representation between indigenous and Western knowledge as polemic categories of knowledge, each being stagnant and essentialized categories. In the end, the promotion of indigenous knowledge systems stumbles into many of the same pitfalls as that criticized of Western scientific knowledge. For example, by labeling a form of knowledge as indigenous knowledge its proponents are defining the boundaries that this system of knowledge encompasses and relegating it into an essentialized and often romanticized category that stands in opposition to Western scientific knowledge.

Even more problematic is that those who label a field of knowledge as indigenous are often not of the indigenous population that they choose to represent and their interests in promoting an indigenous knowledge are sometimes ambiguous. In the case of raised fields, the indigenous knowledge at stake was "discovered" and promoted by non-local elites and North American researchers. The question to be answered is, what were the political and economic interests of advocates of raised field rehabilitation? As DeWalt notes (1999), at the very least researchers should conduct a frank self-appraisal of their own interests in promoting an indigenous knowledge.

In an article by Agrawal (1995), the author argues that the claiming of any form of knowledge as indigenous knowledge likely has more to do with politics and the prestige value of advancing an indigenous knowledge, than it has to do with altruistic attempts to foster economic growth. For example, as I will argue for the case of raised fields, it is no coincidence that interest and research on indigenous knowledge systems in Latin America has often gone hand-in-hand with social movements that advance the rights of indigenous peoples. In fact, Agrawal (1995) agrees that the differences between Western scientific knowledge and other forms of knowledge are mostly cultural, and that science is not a value free, universal methodology, though it is very likely the dominant form of knowledge. In his conclusion, Agrawal calls for more recognition of the politics of knowledge, particularly recognizing who controls the systematizing and dissemination of knowledge, and the assessing of how such knowledge "differentially benefits different social groups" (1995:433).

In the end, political interests have already been advanced through the advocating of indigenous knowledge, since it has located historic political struggles for human rights into academic and development discourses. For any knowledge to be deemed indigenous knowledge, it is implicitly recognizing the history of colonialism that established the West's dominance over indigenous peoples. There could not even be a term such as "indigenous" if there were no prior conquest and domination by the West (Purcell 1998). Under colonialism, therefore, Western scientific knowledge is also recognized as dominant over indigenous knowledge. Thus, by raising the issue of indigenous knowledge as equal and on par with Western scientific knowledge, advocates of indigenous knowledge have made it into a political and ethical issue (Purcell 1998).

These political and ethical issues, such as the empowerment and advocacy of indigenous groups, should be examined and explored when advancing the cause of indigenous knowledge in development. For example in the case of raised field agriculture in Bolivia, the practice of archaeology has been undeniably intertwined with state-sanctioned nationalism (Mamani Condori 1989). This leads into interesting questions about the political symbolism of raised fields as it is tied to state nationalism. On the other hand, another interesting parallel is how the actual Aymara speaking farmers, who are the alleged descendants of the once expansive Tiwanaku polity, view raised fields as a political and cultural symbol. In fact, it was the outside development workers and archaeologists who had a much stronger interest in the cultural heritage of raised fields, than the Aymara speaking farmers who were asked to rehabilitate them and who were the supposed descendents and inheritors of this cultural heritage.

In a recent book by Shepard Krech III (1999), the author makes a persuasive argument about how the image of the Native American Indian is engrained into North American consciousness. In his book, Krech argues that North Americans of European descent hold an image of Native Americas that represents them as having a fundamental difference in the way they think about and relate to land and natural resources. These "ecological Indians" were innate conservationists who used natural resources prudently and understood them ecologically.

I draw upon Krech's (1999) observation about North American pre-conceptions of Native Americans as ecological Indians. I argue that the pre-Hispanic raised fields were regarded as an indigenous knowledge due to preconceptions about indigenous peoples in the Americas as ecological Indians and natural conservationists. The archaeologists and development workers who were rehabilitating raised fields in Bolivia and in Peru, were no doubt influenced by North American environmentalism and the tenets of the new sustainable development paradigm that sought more environmentally friendly forms of economic development. With much enthusiasm on the part of archeologists and development personnel for ecologically sustainable development, they uncritically applied the concept of indigenous to raised fields. The assumption was that this so-called indigenous knowledge was more environmentally sustainable than development projects based on Western knowledge. This is because indigenous peoples, the ecological Indians, have an innate understanding of the land and natural resources. Thus, social science researchers and development workers uncritically extended their own preconceptions about indigenous peoples and indigenous knowledge to include the raised field system. Like the ecological Indians that Krech (1999) describes, the raised fields were also represented as a form of knowledge that was in harmony with nature and the local environment due to their indigenous ancestry.

In the following case study of the rehabilitation of pre-Hispanic raised field agriculture in Bolivia, the raised fields were represented as an "indigenous knowledge." They are considered indigenous to the Lake Titicaca region due to their extensive pre-Hispanic use, though they had all but disappeared as a method of cultivation in the Lake Titicaca Basin by the arrival of the Spaniards. I argue that there is an underlying connection between the concept of the indigenous, and Western preconceptions of the environment and nature that are being portrayed in the representations of raised fields as indigenous knowledge.[2] This link between indigenous and the natural environment will be explored, particularly as it pertains to development, throughout this case study.

THE COMMUNITY OF WANKOLLO

This is a case study of a rural development project that was based on archaeological investigations and other experiments with raised field agriculture. Raised fields are a method of agriculture that was practiced extensively in the pre-Hispanic past of the Lake Titicaca Basin, but had been abandoned at least since the arrival of the Spaniards in the 16th century. My methodology included interviews with archaeologists, development workers, and other researchers who experimented with and promoted raised fields, both in Bolivia and Peru.[3]

In order to understand the economics of agricultural production in the Lake Titicaca Basin, I undertook ethnographic research in a community that had participated in the raised field rehabilitation project in Bolivia. None of the reconstructed raised fields built by Bolivian communities working with the NGO *Fundación Wiñaymarka*[4] were still being cultivating in 1996 when I returned to Bolivia to commence fieldwork. Hence, I was unable to do economic research that directly measured and compared contemporary raised fields with other methods of agriculture. Instead, I conducted quantitative and qualitative research on the economics of agriculture in a Lake Titicaca Basin community that had participated in the raised field rehabilitation project. I compare these data with oral histories from community members about the raised field rehabilitation project, and interviews on raised field production with development workers. I also consulted secondary sources on raised field production provided by both the NGO that conducted the rehabilitation project in Bolivia, and archaeologists who experimented with raised field production.

The community I chose to work in, Wankollo, had participated in the raised field rehabilitation project led by the NGO *Fundación Wiñaymarka*. Various members of the community of Wankollo had worked with the NGO cultivating raised fields for 5 consecutive agricultural seasons from 1989 to 1994. Wankollo had both "community" raised fields, as well as "private" raised fields. The community fields were located on communal land and were built and cultivated by a large group of people drawn from throughout the community. The private raised fields were located on household plots belonging to individuals in the community and were built and cultivated by their extended family and neighbors.

The Blanco family, who live close to the local Catholic University Agricultural Extension Center where raised field experiments were being conducted in the late 1980s, built the first raised fields in Wankollo in 1989. During the following year, the NGO *Fundación Wiñaymarka* began to work with the community as a whole. Wankollo residents built and cultivated the first 1½ hectares of community raised fields in the 1990–91 agricultural season. Inspired by the good harvests in the first season on the community raised fields, and with the continuing financial support and leadership of the NGO, the community built and planted an additional section of raised fields in 1991–1992. Also during this season, three individual families built three additional small plots (¼ hectare each) of private raised fields with the aid of their extended kin and neighbors. In total, there were four private plots of raised fields planted by individual households and two sections of community raised fields built by the community as a whole. All of the raised fields that were built in Wankollo were built with the assistance of development workers from the NGO, and all the participants were given different types of aid, such as foodstuffs, hand tools, and potato seed for participating in the project.

The community of Wankollo is an ex-hacienda. A hacienda is a consolidated private estate that had been owned by a single landowner (usually of Spanish descent) prior to the agrarian reform. Following the agrarian reform in 1953, the laborers of the former hacienda were granted ownership of parcels of land in the community. Wankollo lies on the relatively flat rolling plains of the Tiahuanaco Valley and has no access to the nearby hills that separate the Desaguadero, Tiahuanaco, and Catari Valleys. This is an important aspect of its physical geography, since the nearby hills offer the best protection from the periodic frost that plagues agriculture on the high plains. The Tiahuanaco Valley is located on the high plains of the southern Lake Titicaca Basin, a particularly harsh, cold, and dry environment. Though the community of Wankollo is in the southern Lake Titicaca Basin, it is still far enough away from the lake (about 20 km) that it lacks the lake's direct warming effect (see Figure 2—Map of Bolivia, and Figure 3— Map of the Lake Titicaca Basin).

Wankollo is located southeast of the town of Tiahuanaco and borders the town. The town of Tiahuanaco is the seat of local and regional government, and the hub of regional economy. The nation's capital city of La Paz is less than 70 km from Wankollo and can be easily reached by bus within two hours, particularly since construction of a new paved highway in 1997 that runs through the center of the community and links it directly to La Paz. The Guaqui–La Paz rail line that connects the international port of Guaqui with the capital city of La Paz runs across the northern border of Wankollo. The old dirt road that previously connected Guaqui, Tiahuanaco, and La Paz, runs along the northern side of the community and marks the border with its northern neighbor communities. Historically, Tiahuanaco has always been well connected to the city and other markets, because it falls on this major trade route from La Paz to the port of Guaqui and the international border with Peru. However, the importance of this route has declined a bit in recent decades as trade with Argentina, Brazil, and Chile has increased.

Wankollo also borders the archaeological site of Tiwanaku[5], which is now controlled by the National Institute of Archaeology and Anthropology (DINARA). A large section of the official site of Tiwanaku was once part of the hacienda of Wankollo.[6] The archaeological site at Tiahuanaco was the epicenter of the pre-Hispanic civilization of Tiwanaku.

The three valleys of the southern Lake Titicaca Basin—the Desaguadero, Tiahuanaco, and Catari Valleys—make up what Kolata (1986) describes as the "agricultural heartland" of the pre-Hispanic Tiwanaku polity. Wankollo, therefore, sits squarely at the center of this ancient agricultural heartland.

At one point in colonial history, Wankollo was a corporate Indian community, with clearly marked boundaries and communal title to the land. However, at some point, the community came wholly into the possession of a single estate owner (*hacendado*). The original corporate Indian community members (*originarios*) were forced out of the community prior to the turn of the 20th century. By the time of the agrarian reform, all of the workers (*colonos*) who lived on the hacienda of Wankollo and who worked for the landowner (*hacendado*) were not original members of the community. After the 1953 agrarian reform, a large group of Aymara speaking people, who claimed to be the descendents of members of the original corporate Indian community (calling themselves *ex-comunarios*), returned to Wankollo and successfully lobbied for land grants in the community.

With the return of the *ex-comunarios* to the community of Wankollo after the agrarian reform, households became stratified into two principal categories. On one side of the community are the *ex-comunarios*, who had been driven from Wankollo by the *hacendado* and who had returned in large number to reclaim their land rights following the revolution. The second group consisted of the workers (*ex-colonos*) who had held usufruct rights to land in Wankollo and had worked for the *hacendado* immediately prior to the revolution. These two groups both received land under the agrarian reform of 1953, but they continue to remain hostile to each other even to this day. The groups are separated physically, socially, and politically within the community. Physically, the two groups are literally located in two different sections of the community. The *ex-comunarios* were granted land to the west of the stream that runs through the community. This is the half of the community closest to the town of Tiahuanaco and the pre-Hispanic site of Tiwanaku. The *ex-colonos* received their land in the eastern half of the community, farther from town (see Figure 4— Agrarian Reform Map of Wankollo).

There is also a significant social distance between the two groups. Part of this is linked to the physical locations of the two groups and their household plots of land. For example, though the average household plot for *ex-comunarios* was smaller than that of the *ex-colonos*, their households were located on the western side of the community closest to the town of Tiahuanaco. Therefore, in general the *ex-comunario* population has been able to better take advantage of the resources and economic relationships with the town of Tiahuanaco and the tourism generated by the site of Tiwanaku. These *ex-comunario* households were also somewhat more accustomed to dealing with the state and participating in the national and regional economy. While the *ex-colonos* had previously dealt primarily only with the landlord, many of the *ex-comunarios* were returning Chaco War veterans who had been living in the cities and other regions of Bolivia for several generations. These *ex-comunarios* had a better command of the Spanish language and more experience dealing with Spanish culture and taking advantage of economic opportunities that were available in the town. Even today, these households are more likely to have children in the local high school in Tiahuanaco and to participate in the tourist economy generated by the site of Tiwanaku and other regional archaeology projects.

The political separation between these two groups in Wankollo is also evident. For example, the community is organized politically as a peasant union (*sindicato*) with elections held every year for the several community posts. The highest post is *secretario general*. This post is supposed to rotate every year between an *ex-comunario* and an *ex-colono*. However, an extended family of *ex-comunarios* has been able to dominate these posts and other economic activities in the community. For example, the *secretario general* of the community in 1996–97, Jose Condori, was originally from another nearby community. After his marriage to the daughter of a powerful *ex-comunario* who had been influential in the agrarian reform process in the region, Jose bought land in Wankollo and quickly became active in community politics with the aid of his powerful new in-laws. Despite rules to the contrary, Jose was elected to an unprecedented second consecutive term in office. During this term his brother-in-law became my own field assistant, his access to this lucrative job as my field assistant no doubt linked to his family connections.

In recent years, the community of Wankollo has been something of a leader amongst the surrounding communities for trying new projects and experimenting with new development plans. The members of the community, especially the younger households, are usually willing to listen to development groups talk about new technologies and new experiments in agriculture. The many buildings and activities going on in the community attest to this. Aside from the usual elementary school house and community building, Wankollo also has a community chapel, a community pigsty that is run by several families together, a large community greenhouse that grows crops of lettuce and tomatoes for community consumption, and a natural medicine hospital that services the surrounding communities.[7] Wankollo does not have electricity because the community is quite large with houses spaced far apart from each other. The cost of extending electricity across the whole community is prohibitive and even with the local development aid that has been offered to assist in this project, it is estimated that it will cost US$200 per household. Most other communities in the area have electricity. However, another earlier rural development project helped to install solar panels in a small number of homes, which generate some electricity for those particular households. Other indicators of wealth in the community include a number of households with trucks and motorcycles, and one household that even has a tractor. No households in Wankollo have running water and all rely on household wells for their water, though the water table is generally very close to the surface in this region.

Community activities usually take place in the courtyard of the natural medicine hospital and in the community buildings that are a part of the hospital complex. The yearly community festivities take place in the plaza area between the chapel and the hospital. This is where the bands play and the dancers perform. Along the plaza, venders come and set up kiosks selling beer and other beverages.

Aymara is the primary language of the community and is spoken in households and in all community meetings. Spanish is more widely spoken by the younger generations, particularly by men in their 20s, 30s, and early 40s. The *ex-comunario* households generally have more proficiency in Spanish, including the members of the older generations. However, most elders and very small children are monolingual Aymara speakers.

There were 90 households in Wankollo and a population of 400 people in 1996. However, females outnumber males across all age categories due to male out-migration. For example, one out of every four households in Wankollo does not have any adult males permanently residing in it. In those households that do have male residents, often the men are away from their homes for most of the week or month working in other locations. In situations where men are absent from the household, women are in charge of the day-to-day agricultural activities, including the harvests.

No households in Wankollo produce enough potatoes to meet all of their yearly consumption needs on a regular basis. Most households consume what potatoes they do produce during the period of time between the harvest and the next year's sowing. During the sowing, most households plant what potatoes they have left as crop seed and often must buy additional potato seed to augment their dwindling supplies. After the planting season, nearly all households have to buy potatoes for household consumption until the next harvest comes in.

The high plains of the Lake Titicaca Basin is a cold frost prone region, and the only staple food crops that can be grown are potatoes and other native tubers, *habas* (fava beans), and *quinua* (another native seed crop). There is little market for these native crops, except potatoes. However, most households on the high plains primarily cultivate varieties of potatoes that have very little market value, because they are either too small for commercial sale or are a bitter variety of potato. Households choose to plant these non-marketable varieties of potato because they are more frost resistant than the typical "sweet" potatoes that are sold in the cities and are familiar to North Americans.

Because crops are not grown for the market, all households rely on off-farm income. This income may include wage labor in the nearby towns and in the city of La Paz. Many men migrate regularly in search of wage labor; some even work Mondays through Fridays in the city and return only on the weekends to be with their families. Other community members seasonally migrate to farms at lower altitudes, particularly the yungas region on the eastern slopes of the Andes. The importance of seasonal migration and the impact it has on the availability of labor for intensive agriculture in the community of Wankollo will be discussed in Chapter 7.

OUTLINE OF CHAPTERS

In Chapter 2, I present a brief introduction to the social history of the Lake Titicaca Basin and Bolivian high plains (the altiplano). Archaeologists have associated the pre-Hispanic raised fields of the southern Lake Titicaca Basin with the ancient Tiwanaku polity. However, perhaps even more importantly, the raised field rehabilitation project symbolically linked raised fields to the ancient Tiwanaku polity. Thus, I chose to begin the chapter with an introduction to the social, political, and economic organization of this pre-Hispanic civilization. An important part of chapter 2 is highlighting how the Aymara groups who have lived in the Lake Titicaca Basin, at least since pre-Incaic and pre-Hispanic periods, have adapted to and exploited their social, political, and physical world. In particular, I pay special attention to the ways local Aymara groups interacted with the state from the first Inca invasions, through the Colonial period, the post-independence Republic of Bolivia, and into the 20th century. This chapter shows how the local and regional political economy has changed, and changed drastically, since the time when raised fields were once flourishing and prolific in the Lake Titicaca Basin.

Chapter 3 examines the intersecting social, political, and economic trends developing in the latter half of the 20th century that made the raised field rehabilitation project so appealing to both the Bolivians who promoted and supported the project, and the international development agencies that funded the project. Raised field agriculture first gained the attention of researchers and academics, followed by development groups in Bolivia and international funding agencies, and eventually the news media in Bolivia. What were the social, political, and economic circumstances at the time of the raised field experiments and subsequent development project in Bolivia? Why did the project so quickly and easily win the attention and imaginations of researchers, development personnel, and eventually the public media in Bolivia?

To answer these questions, I investigate three interrelated trends leading up to the implementation of the development project and the re-introduction of raised fields back into communities in the Lake Titicaca Basin. First, I take a look at political and economic trends in Bolivia following the Agrarian Reform and through the neo-liberal economic policies of the mid-1980s. Second, I explore the pervasiveness and application of ethnic claims in nationalist and indigenous political movements in La Paz. And third, I critically review the history of development and the turn towards environmentally friendly projects via the "sustainable development" model of international aid and assistance.

Chapter 4 is a key chapter that critically explores how the raised field rehabilitation project in Bolivia, and raised fields in general, were represented by academics and development industry workers, particularly through an analysis of the written publications produced by these groups. I begin the chapter with a review of the archaeology of raised field agriculture and the Lake Titicaca Basin. I then turn to the specific implementation of the raised field rehabilitation project in Bolivia beginning with the first initial experiments on raised fields in Bolivia in the mid-1980s and following along the course of the rehabilitation project as it grew in size and scope to include over 50 communities and 90 hectares of land.

After this introduction to the project, I delve into the representation of raised fields, particularly those representations produced by the raised field rehabilitation project in Bolivia. I frame this analysis of the representations of raised fields into three categories or themes; raised fields as "indigenous knowledge," raised fields as "ecologically sustainable," and raised fields as "appropriate technology." I unravel their history, and critically explore the meanings of these three terms, demonstrating how each was applied to the raised field rehabilitation project. I elucidate some of the assumptions about local agriculture and local peoples in the Lake Titicaca Basin that lie within these representations of the raised fields. The representations of the raised fields are the key to why this method of agriculture was so appealing to development workers. However, as I will show in the following chapters 5 and 6, the representations of raised fields clearly made erroneous assumptions about the contemporary social and economic contexts of agriculture in the Lake Titicaca Basin, such as access to land and labor for agriculture.

In chapter 5, I review each of the representations of raised fields in light of local agricultural practices and the economics of agriculture in the Bolivian Lake Titicaca Basin. I demonstrate discrepancies between the representations of the fields and the actual contemporary practice of farming on raised field beds. For example, raised fields were regarded as ecologically sustainable primarily because they were supposed to be capable of continual cropping. However, the raised fields in Bolivia were not continuously cropped and were abandoned after 2 to 4 years of cultivation, because the fields had lost their fertility and needed to be returned to fallow. I also examine concerns that raised field agriculture might not be compatible with local patterns of land tenure and general land scarcity in the high Andes. However, I reject these as significant limiting factors for contemporary usage of raised fields because I demonstrate that raised fields did not compete with other types of agriculture for access to land and that land pressure in general has decreased in recent decades.

In Chapter 6 I take up the issue of access to labor for agriculture. My goal is to demonstrate that access to labor for raised fields that was the primary obstacle for building and cultivating the raised fields in contemporary Bolivian communities. The raised fields required large investments of labor for their construction, and given that the fields were not replanted annually in potatoes, this large initial labor investment was not returned with adequate production. I argue that researchers have overemphasized the production per unit of land on raised fields, rather than production per unit of labor. As I demonstrate, it is labor that is the scarcest resource for agricultural production in Wankollo. Therefore, researchers should take much more into account the availability of labor when building models of agricultural production on raised fields. I conclude that production per unit of labor is actually higher on regular dry fields than it is on raised fields.

Finally, in chapter 7 I discuss the raised field rehabilitation project and some of the major implications of this work. I maintain that raised fields are an "invented tradition" by archaeologists and development workers, aimed at garnering academic and development support. I put forward that one reason why raised fields were so appealing to North Americans and upper and middle class Bolivians of Spanish descent, has to do with deeply held preconceptions about indigenous peoples as natural conservationists and ecological Indians. I also suggest that the raised field rehabilitation project was upholding an ideal of an "indigenous peasant farmer" far removed from the actual lives and conditions of the local peoples living in the Lake Titicaca Basin region. I argue that the Western academics and upper class Bolivian development workers who promoted raised fields are preserving these idealized images of the indigenous peasant cultivator. By rehabilitating an ancient indigenous knowledge, both international and national development aid workers are marking a boundary between themselves and the rural Aymara speaking farmers of this region. Paradoxically, while mestizo Bolivians[8] are at once distancing themselves from rural indigenous farmers, they are also simultaneously claiming this indigenous knowledge as a part of their own cultural heritage.

NOTES

[1] The floating gardens of the Aztecs were somewhat different in form and function than the high altitude fields of the Lake Titicaca Basin.

[2] I will continue this discussion of the link between the raised fields as indigenous knowledge and Western preconceptions of the indigenous peoples and the environment in the concluding chapter of this book.

[3] See Appendix 1 for a more detailed description of methodology.

[4] The non-governmental organization (NGO) *Fundación Wiñaymarka* was headed by the then director of the Bolivian National Institute of Archaeology who had been involved in the initial experiments with raised fields in Bolivia in the mid-1980s. The NGO was formed with the objective of "recovering" Andean culture through the rehabilitation of raised fields and other forms of indigenous agriculture.

[5] I use the Spanish spelling *Tiahuanaco* to refer to the modern town and the valley, and the Aymara spelling *Tiwanaku* to refer to the archaeological site and the pre-Hispanic civilization.

[6] Both community members from Wankollo and archaeologists at the National Institute of Archaeology confirm that the hacienda of Wankollo once owned land that is now controlled by the DINARA. See also Figure 4—Agrarian Reform Map of Wankollo and note the top left hand side, shows two parcels that do not have any numbers assigned to them. This is because these parcels were not allocated to community members, and are controlled by DINARA.

[7] The herbal and natural remedies practiced at the hospital have been greatly reduced in recent years, and I rarely saw any patients staying at the hospital or visiting the hospital staff while I was in Wankollo.

[8] Mestizos are persons of mixed Spanish and indigenous background.

CHAPTER TWO

Ethnic Groups and the State: From Tiwanaku to National Revolution in the Lake Titicaca Basin

OURISTS FROM EUROPE AND NORTH AMERICA COME DAILY TO THE town of Tiahuanaco making the one and a half-hour trip (71 km) from the capital city of La Paz, Bolivia. Many are traveling from north to south on their trek through the Andes and have already visited other imposing pre-Hispanic sites in Peru, such as the Inca fortress of Sacsahuaman at Cuzco and the magnificent towering Inca site at Macchu Picchu. They come to the town to visit the archaeological site of Tiwanaku and along with them come thousands of Bolivians, who visit the site annually, particularly the busloads of school children who are taught about this archaeological site and the ancient Tiwanaku civilization. The tourists are led on tours through the site of Tiwanaku where tour guides speak of an ancient civilization that built this magnificent city and ceremonial center. They explain that Tiwanaku was once the capital of an ancient Andean state whose epicenter was on the Bolivian highlands and that preceded the Inca Empire centered in the Peruvian highlands by nearly 1000 years. The tour guides point out various pre-Hispanic Andean symbols at the site, such as the "Andean Cross" (see Figure 5—Andean Cross) and the "Staff God" (see Figure 6—Staff God) and relate them to primordial Andean cosmology. They also talk about the great technology found in the construction of the ceremonial monuments at the site and of the complex organization of raised bed agricultural fields that supported large numbers of people on the high plains in the pre-Hispanic past. Through the oral representations given by the tour guides, the site and ancient civilization comes to life again, if only in the minds and imaginations of the tourists. But Tiwanaku has special meanings for the Bolivians who visit the site. For them, the site is reified as a symbol of Bolivian cultural history, heritage, and identity, and it stands as a monument toward their future as a nation and as a people.

The archaeological site of Tiwanaku is a multivocalic symbol important in the identity of Bolivian people. However, the symbolism of Tiwanaku carries different meanings for different segments of Bolivian society. The middle and upper class La Paz school children that visit the site are taught about the ancient city and civilization in terms of nationalist identity. For them, the former civilization whose capital was once located at the site of Tiwanaku is both a symbol of the Bolivian past and also a symbol of Bolivia's future. As the former director of the Bolivian Institute of Archaeology Oswaldo Rivera claimed, "Tiwanaku is a reminder to Bolivians that their past glory can also be their present and their future" (cited in Sagarnaga 1991). According to state sponsored archaeology in Bolivia, Tiwanaku was the capital of a far-reaching and politically influential pre-Hispanic state whose heartland was in the southern Lake Titicaca Basin that is now present-day Bolivia. It was a multi-ethnic, class-based society with landless members, specialized craftspersons, commoner agriculturalists, and elites. This is an integral part of the meaning embedded in Tiwanaku as a symbol of Bolivian nationalism. The future that Tiwanaku is symbolizing on this nationalist agenda is of a Bolivian state that co-opts the multiple ethnic origins of its many indigenous peoples, often using rhetoric similar to *mestizaje*[1] in the political discourses of other Latin American countries.

Of course, school children are not the only Bolivian visitors to the site of Tiwanaku. Beginning in the 1990s, indigenous leaders from the local and urban Aymara community began to return to Tiwanaku to reenact their own version of Aymara renewal in a celebration of the Aymara New Year each June 21st during the winter solstice. For them, Tiwanaku is a symbol of Aymara indigenous ethnic identity and in some instances it is employed as a powerful symbol for a new generation of urban Aymara residents who continue to vie for political power at the municipal and state level in La Paz. Indeed, the Tiwanaku polity was multi-ethnic, and many believe the Aymara are the direct descendants of this once powerful pre-Hispanic polity (Browman 1994). Through the claiming of Tiwanaku as a political symbol, Aymara leaders are reasserting Aymara political power; a power rooted in the past when the Aymara were in a position of political dominance in the Lake Titicaca Basin.

Stern (1987) uses the terms "nationalist" and "nativist" to label these two versions of Andean political strategies. The nationalist version is represented in the symbol of Tiwanaku as portrayed by state-sponsored archaeology, while the nativist version lies in the Aymara ethnic claims on Tiwanaku as a symbol of emergent indigenous identity. Stern (1987) claims that there is a historic tension between these two poles of revolutionary Andean politics. Nationalistic movements, in general, often use of images of the past as a basis for contemporary claims to national unity of a specific geographic area and the ethnic groups that fall within its socially constructed boundaries. It is of no small consequence that the heartland of the Tiwanaku polity is wholly within present day Bolivian boundaries, while the capital city of La Paz lies just 71 km from the 1000–year-old capital of the ancient Tiwanaku civilization.[2]

The image of Tiwanaku as an expansive state holding political economic influence and power outside of the current boundaries of Bolivia—and into the territories of present-day Peru and Chile—also serves as a rallying point for contemporary Bolivian nationalism. For example, see Figure 7—The Tiwanaku State. This illustration is of a map of the expansive pre-Hispanic state as it is depicted in a pamphlet on raised field reconstruction produced by the NGO *Fundación Wiñaymarka*.[3] The map draws the boundaries of the pre-Hispanic Tiwanaku state to include the current politically contested coastal areas of Peru and Chile. This is a clear example of contemporary political and nationalistic claims that have been expressed in the symbol of Tiwanaku as an expansive pre-Hispanic state.

In opposition to the nationalist political symbolism of Tiwanaku, Aymara groups also appropriate Tiwanaku as a symbol of "nativist" or indigenous political unity. For emergent Aymara speaking ethnic groups, it symbolizes the specific political claims of this subaltern population. Through the symbol of Tiwanaku, these Aymara groups are expressing a shared indigenous group membership. However, although the symbol of Tiwanaku for indigenous groups is an ethnic symbol meant to unite an ethnic group, it also has significant links to class issues. This is because the appropriation of Tiwanaku as a political symbol is by a specific class of urban Aymara who are paradoxically educated yet still of the working class, and who have less access to economic opportunities than middle and upper class mestizos. That this symbolic appropriation of Tiwanaku is linked to newly urban Aymara, and not necessarily to their rural agrarian kinsmen, is significant and reveals the class issues behind this so-called ethnic symbol. It is no coincidence that social movements drawing on ethnic claims to a pre-Hispanic past thrive in the capital city, but have less impact in the contexts of rural life, even within the town of Tiahuanaco itself.

This chapter traces the nationalist and the nativist/indigenous precursors that inform the contexts of contemporary political struggles between indigenous groups and the state in Bolivia. Nationalist movements that use ethnic claims to the past first surfaced in the Andes in the early 20th century in the form of *indigenismo*[4] in Mexico and Peru (Kristal 1991), and to a lesser extent in Bolivia. Nationalist political movements in Bolivia gained momentum following the defeat of Bolivia in the Chaco War (1932–1935) with populist political pacts and worker-peasant movements that climaxed in the 1952 victory of the National Revolutionary Movement (MNR)(Klein 1992). On the other hand, nativist/indigenous political stances can be traced at least as far back as the rhetoric of indige-nous peasant rebellions of the late 18th century, and in the case of the Bolivian Aymara of the Lake Titicaca Basin to the Túpac Katari Rebellion of 1780–82 (Valle de Siles 1990). It is important to review the historical precedents of current political movements, such as the Bolivian National Revolution, the Túpac Katari Rebellion, and the Tiwanaku civilization to explore the foundations of the collective memories that both current nationalist and nativist political movements draw upon.

These historical memories shape the contexts of contemporary eth-nic and class negotiations with the state, and it is their symbolic use as rep-resentations of Andean historic consciousness that has the ability to spur modern political action. Through both nationalist and nativist representa-tions of Tiwanaku, this contested symbol reproduces a historically situated social struggle between ethnic and class groups. For the nativist social movements, their contemporary leadership is drawn from a newly urban-ized group of Aymara who have only recently left behind their agrarian lifestyle and are taking part in the urban working class.

In this chapter, I focus on the Aymara of the southern Lake Titicaca Basin and the manifold ways they have resisted political domination and shaped the economic world that they have lived in through more then a millennium of occupation on the high plains of Bolivia. I establish the wax-ing and waning of political influence and power in the region, as well as trends in economic organization and subsistence strategies that character-ized the lake basin landscape from the height of Tiwanaku's political power to the 20th century and the Bolivian National Revolution. Throughout this chapter, I focus on the engagement of ethnic groups with the state, specifi-cally focusing on the Lake Titicaca Basin. My objective is *not* to give an exhaustive review of the history of the lake basin from Tiwanaku to the present, but to focus on the strategies employed through time by ethnic groups and economic classes as they resist political domination, participate in new economic situations, and continually redefine their social identity. Throughout this section I touch on key moments in the history of the region that ignite the "moral memory"[5] of the people living in the Lake

Titicaca Basin and on the high plains of Bolivia, who participate politically, socially, and economically in the modern nation-state of Bolivia.

THE PHYSICAL SETTING

The Tiahuanaco Valley lies on the high plains of the Bolivian *altiplano*,[6] off
the southern tip of Lake Titicaca, the world's highest navigable lake (3810
meters above sea level). The altiplano and the lake are nestled between two
tall mountain chains, the Cordilleras Occidental and Oriental, which form
part of the high Andes Mountains that run the length of the western coast
of South America. The central Andean altiplano stretches for over 2000 km
from north to south, measuring nearly 200 km at its widest point from east
to west. The elevation of the Basin floor varies from about 4500 m in the
north, decreasing to 3600 m at its southern end. The surrounding mountain ranges of the Occidental and Oriental reach over 6000 m while the
mountains bordering Lake Titicaca vary between 4500 m to 5400 m, with
the sole exception of Nevado Illampu, the tallest peak in the altiplano basin
area at 6338 m.

From this lofty perch, there are three valleys that formed the "agricultural heartland" of the Tiwanaku polity and which are located within
the modern state of Bolivia, comprising one of the most densely populated
rural regions of Bolivia today. These valleys include the Catari Valley to the
north and east, the middle Tiahuanaco Valley that was home to the ancient
city of Tiwanaku, and the Desaguadero Valley to the south and west of
Tiahuanaco. One might imagine the altiplano to be an endless flat landscape, but in fact it has its own mosaic of varied microclimates. These three
valleys in the southern lake basin are separated by inter-valley hills, and
each valley has a distinct climate ranging from the wetter and warmer
Catari Valley to the dryer, colder Desaguadero Valley. Likewise, the basin
floor itself rolls and dips so that during the rainy season there are distinct
patches of land with some areas inundated by water and other areas above
the water line.

The climate of the Lake Titicaca Basin and northern Bolivian altiplano is marked by distinct seasons of wet summer and dry winter, as well
as extreme diurnal temperature fluctuations. This results in a short growing season with a strong tendency towards frost at any point during the
agricultural cycle. Certain variants of the primary native Andean crop, the
potato, have adapted to this extreme environment and can withstand some
amount of frost. For example certain "bitter" variants of potatoes resist
frost better than the white "sweet" varieties like those grown in the United
States. Maize does not grow at this altitude except for in a small number
of sheltered locations on or near the lake. As we will see in the following

pages, the lake basin was once a very productive agropastoral zone, though the products available for production were limited to a few well-adapted cultivars and animal species. The fact that these products are less valued in the modern Bolivian economy has much to do with the area's current declining productivity and will be explored further in later chapters of this book.

THE TIWANAKU STATE AND POLITICAL ECONOMY

The pre-Hispanic epicenter of the Tiwanaku polity was located at the southern end of the Lake Titicaca Basin in the present day Tiahuanaco Valley. After A.D. 400, Tiwanaku began to extend its political and economic power outside of the local core area in the southern lake basin. Tiwanaku's influence expanded into far off regions such as the tropical Bolivian *yungas* region to the east, the Moquegua Valley to the west, and the Cochabamba Valley to the southeast. During the peak of Tiwanaku influence (about A.D. 700–1000), it was the center of social, economic, and political power in the south central Andes.

Disagreement over the political organization of Tiwanaku continues, however, Kolata (1993) describes Tiwanaku as an expansive state that employed a "mosaic of strategic policies and political relationships" linking diverse populations and ethnic groups. One such strategy employed by the state was through direct colonization of distant outposts. Kolata argues that colonists from the Tiwanaku core area colonized far off areas and formed distinct ethnic enclaves within populations in foreign territories. Tiwanaku colonists retained their identity and remained separated from the locals as characterized in their distinct clothing, goods, and other aspects of material culture (Janusek 1994; Kolata 1993). For example in the case of the Moquegua Valley, according to Goldstein (1989) Tiwanaku held directly controlled colonial settlements that eventually led to the disappearance of the local ceramic tradition. Kolata (1993) argues that this political strategy of colonization of key areas was the foundation of state political power outside of the core region and managed the control of key economic resources. This model is similar to the "vertical archipelago" argued by Murra (1968) for the later Lupaqa kingdom of the Aymara that preceded Inca conquest.

The economy of the Tiwanaku polity was agropastoral and relied on the intensive cultivation of potatoes and other native plants, in combination with the high altitude herding of large llama and alpaca herds. Kolata (1993) argues that the most important system of production was the intensive cultivation of low-laying, marshlands through raised field agriculture. The raised fields in the Lake Titicaca Basin during the

Tiwanaku era were elevated platform beds ranging from 5 to 10 meters wide and up to 200 meters long. The raised beds were surrounded by long and deep canals, which caught and retained ground water.

Questions remain about the pre-Hispanic organization of labor for cultivation of raised fields and the level of integration of raised field agriculture with state control. Kolata (1993) argues that state administrators, who employed rotational workers made available through a system of corvée labor, managed the agricultural heartland of raised fields, particularly in the Catari Valley. There is little doubt that building such a large expanse of raised fields represented an intense investment of labor and planning. However, it is contested whether raised fields were necessarily managed and controlled by the state (Erickson 1993, 1999). Less in doubt is the surplus agricultural goods that these raised fields produced, which was able to maintain a large urban population on the altiplano and provided the bulk of subsistence needs. Kolata (1993) contends that it was control over this fundamental resource that was the basis for elite power in pre-Hispanic Tiwanaku society.

Yet there were many agricultural products and goods that were unavailable at the high altitude of the Lake Titicaca Basin due to the cold environment. Particularly important to pre-Hispanic states would have been access to maize and coca. While it seems clear that Tiwanaku could have met its own basic subsistence needs, both maize and coca were important for ritual uses. Coca was chewed as a stimulant and maize was brewed into maize beer, known as *chicha*. Both coca and chicha were probably important for maintaining an economy based on reciprocity and redistribution by Tiwanaku elites and their client groups (Kolata 1993). Therefore, to complement and supplement the economy of intensive agriculture and herding, residents of the Tiwanaku core may have maintained direct rights to land at lower elevations or ecological zones. Maintaining plots of land in different locations and ecological zones is termed "ecological complementarity" and it is an economic strategy whose practice in the southern Andes has been long-standing.

When distance away from the lake basin was more than a few days in travel time, the Tiwanaku polity maintained ethnic enclaves in a colonialist relationship with the capital of Tiwanaku, as described previously in the vertical archipelago model (Kolata 1993). These ethnic enclaves were apparently interspersed with the native population in these far off regions such as the Moquegua Valley. The terms "verticality" and "ecological complementarity" have been used to describe this strategy of accessing goods and resources available only in other ecological niches. For the pre-Hispanic polities located in the Titicaca Basin this meant maintaining

access to lower altitudes both to the east and west of the Lake Titicaca
Basin. This vertical exploitation of different ecological zones by a single
ethnic group or community is a system that has a very long history with
residents in the Lake Titicaca Basin and other areas of the central Andes.
Access to different zones was maintained with a variety of strategies, from
the migratory farming of different zones by the community members themselves, by permanent settlements of members from highland ethnic groups
in distant enclaves, or by various indirect trading mechanisms. The
Tiwanaku polity primarily used colonization as a means to maintain direct
access to distant goods and resources, strategically targeting prime agricultural lands for the cultivation of maize and coca (Kolata 1993).

At some point between AD 1000–1100, the site of Tiwanaku went
into decline. This political disintegration coincided with a four century long
drought that commenced at about AD 1000. Since Tiwanaku economic
well being depended on intensive agriculture, it is argued that the deterio-
ration of agriculture throughout the polity, and the ultimate abandonment
of the majority of raised fields, directly caused its political decline (Binford
et al. 1997; Ortloff and Kolata 1993). Ortloff and Kolata (1993) argue that
it was "climatic change in the form of persistent lowered precipitation"
that caused the failure of the agricultural base on which the state depended. The authors maintain that a complex process of social and political factors over a few generations would be necessary for the total collapse of a
state-level political system, unfortunately these factors are generally
unrecordable in the material remains of the archaeological record.
Regardless of social and political factors that contributed to political
demise, in the end the proximate cause of the collapse of the Tiwanaku
polity was the post-AD 1000 drought that severely altered the regional,
and perhaps global climate, undermining the economic base that was the
foundation of the state itself (Binford et al. 1997; Kolata and Ortloff
1993).

Whether or not agricultural decline and the abandonment of the
majority of the raised fields caused political decline or merely coincided
with more complex political and social forces (Erickson 1999), it is evident
that the Tiwanaku polity was defunct by AD 1000–1100. After AD 1000,
life on the altiplano changed significantly with the depopulation of urban
centers and shifts in settlements to dispersed villages and hamlets. The
urban capital of Tiwanaku was practically abandoned at this time, as well
as secondary urban centers in the lake basin (Ortloff and Kolata 1993).
Some archaeologists contend that raised fields were not completely abandoned and that political power merely retracted to local-level interests.
Graffam (1992) argues that power was probably reinvested into local kin-

ship based organization, such as the Andean *ayllu*, and was formed from relationships of common descent. He contends that the concomitant economic and agricultural reorganization that took place following political collapse was redirected into the pastoral economy, with camelid herding being prioritized as the economic basis of these post-Tiwanaku kinship based polities. These post-Tiwanaku kinship organized polities, with their economic emphasis on llama and alpaca herding, are called the Aymara Kingdoms in the ethnohistoric literature.

THE QUESTION OF TIWANAKU ETHNIC ORIGINS

It is important to address the question of whether Tiwanaku was an Aymara polity or whether the Aymara were later arrivals into the lake basin following the Tiwanaku civilization and the political void that was left in its wake. This question is important to investigate in light of contemporary representations of Tiwanaku as an Aymara state, with the contemporary inhabitants of the lake basin the alleged descendants of that state. Undoubtedly, the Aymara were well established in the Lake Titicaca Basin by the time of the Inca conquest since it is well known that the Aymara Kingdoms put up a good fight against the Inca conquerors (Klein 1992; Murra 1986). But what does the linguistic and archaeological record tell us about the ethnic composition of the altiplano prior to that time? Specifically, from what language group were the leaders of the Tiwanaku polity? Were the inhabitants of the lake basin during the Tiwanaku period Aymara speakers? Obviously, given the loaded symbolism of Tiwanaku in the present, this question is a sensitive one to ask, with the answer having contemporary political consequences.

Some scholars hold that the Aymara are more recent arrivals on the altiplano, arriving sometime after the fall of the Tiwanaku polity (Espinoza 1980; Hardman 1985; Torero 1992). These researchers argue that the inhabitants of Tiwanaku and the lake basin were probably Pukina or Uru speaking people. For example, the colonial records refer to a number of different ethnic or linguistic groups in southern Peru, western Bolivia, and northern Chile after the Spanish conquest along with the Uru and Pukina, such as the Chipaya, Urukilla, Changos, and Camanchaca. If it can be proved that these labels are different names for pockets of the same language, than it would support the argument that they comprised an earlier politically and economically integrated population that predate the Aymara kingdoms. In support of this argument, colonial Jesuit priests were instructed to learn Pukina as one of the three major pre-Hispanic highland Andean languages along with Aymara and Quechua (Browman 1994).

Yet it is not clear that Pukina was actually very widely spoken in the Andes, with Colonial records suggesting that it was not (Browman 1994). Further, linguistic evidence suggests that Pukina and Uru cannot be lumped together as dialects of the same language. That the usage of these two terms overlaps, may be because the term "Uru" was used in a derogatory way by colonial Aymara speakers to refer to various small pockets of ethnic groups who were subordinate to the Aymara (Wachtel 1986).

Many researchers, as well as many Aymara themselves, believe that the Aymara are the direct descendents of Tiwanaku. Browman (1994) makes a strong case that Tiwanaku was a multi-ethnic polity and that following its demise, local elites were able to retain their resources rather than remitting them or directing them towards the core at Tiwanaku. Based on archaeological and linguistic evidence, Browman (1994) argues that Tiwanaku was an Aymara state and that Aymara dominance of the Lake Titicaca Basin and Bolivian altiplano had existed for at least a millennium prior to the arrival of the Inca. Further, he posits that the political organization of Tiwanaku was based on a dispersion of discrete Aymara communities throughout the southern Andes region such as the vertical archipelago model of colonization would suggest. Such political and economic organization would result in pockets of Aymara and other language groups integrated into multi-ethnic and multi-linguistic communities throughout the southern Andes as described in the ethnohistoric record of the Aymara kingdoms. In the end, regardless of when Aymara speaking people actually entered the Lake Titicaca Basin, it is the contemporary Aymara who have already appropriated the Tiwanaku past.

THE AYMARA KINGDOMS

The Aymara Kingdoms are the first polities in the Lake Titicaca Basin for which researchers have ethnohistoric data. As discussed above, following the decline of the Tiwanaku polity life in the Lake Titicaca Basin changed dramatically. Not only were there changes in where people lived, but also how they made a living. The political decline of Tiwanaku coincided with a reduced emphasis on raised field agriculture. Settlements during this period shifted away from lakeside habitations associated with this intensive agriculture to higher altitude hilltop sites, with some sites located directly on top of previous raised fields (Stanish 1994). These sites were usually fortified with defensive walls, suggesting an increased level of aggressive military activities in the area. Likewise, the sites were associated with a large amount of corrals for llama and alpaca herding. It seems that the intensification of pastoralism was one response to the decline of raised field agriculture (Graffam 1990; Stanish 1994).

Following Tiwanaku, there developed a number of smaller Aymara polities that dominated the lake basin and southern Andean highlands until the Inca conquest in the late 15th century. Like other Andean societies, the Aymara polities were organized by kinship relationships based on *ayllu* membership. An *ayllu* is a term that has many meanings in the Andes but in this usage it can be conceptualized as a collection of lineages that are connected through kinship, has corporate access to land that is managed by the *ayllu* collective, and claim descent from a founding ancestor. Each *ayllu* was administered by a *jilakata* with *ayllu* lands probably administered in a corporate and communal fashion (Klein 1992).

Each Aymara Kingdom could be conceptualized as the largest organizational unit of the polity and was organized like an Andean *ayllu*, being divided into two separate spheres with an upper and lower half called the *urcusuyu* and the *umasuyu*. Each of the two halves had its own separate leader and controlled separate territories. The *urcusuyu* had its lands oriented in the western highlands and had economic ties in the western valleys, while the *umasuyu* division was usually to the east with economic ties in the eastern valleys. Each polity was made up of a nested hierarchy of *ayllus*, with each *ayllu* divided hierarchically into moieties made up of an upper (*hanansaya*) and lower (*urinsaya*) moiety. The moieties were comprised of hierarchically arranged lineages (Klein 1992).

In addition to the *ayllu* system of organization, their existed regional leaders called *kurakas*[7] who held land independent of the *ayllus* and extracted labor from the *ayllus*. It is unclear whether the status of the *kuraka* was dependent on the favor of a king or represented a class distinction from the *ayllu* members. There also existed a number of personal retainers, known during the Inca reign as *yanaconas*, who worked as serfs or slaves in the households of nobility and regional elites (Klein 1992).

By the late 14th century the Aymara dominated the Lake Titicaca Basin and most of the altiplano from southern Peru and south. The two largest kingdoms of Aymara speakers were the Lupaqa and Colla, who lived on the western and northern shores of Lake Titicaca. Based on ethnohistoric accounts, both Klein (1992) and Murra (1968) maintain that the Aymara polities utilized the archipelago model of direct colonization of lower ecological zones that created islands of ethnic enclaves. According to Klein (1992), both the *kurakas* and the *ayllus* practiced vertical colonization and held direct control of farmland in lower temperate and semi-tropical valleys to the west and east of the altiplano heartland of the Aymara Kingdoms.

However, archaeological research in the Moquegua Valley by Stanish (1992) points to a more fluid model of interzonal economic exchange that emphasizes indirect exchange, probably based on kin-based and non-kin barter exchange, regional trade networks, and elite alliances. Stanish (1992) argues that both direct colonization and indirect mechanisms of exchange are consistent means of accessing different vertical ecological zones by highland polities. He maintains that the influence of the Lupaqa Aymara Kingdom in Moquegua and Inca administration occurred simultaneously, following the arrival of the Inca into the Lake Titicaca Basin. Stanish argues that prior to the Inca expansion into the Lake Titicaca Basin, relationships between the highland Aymara polities and the polities of the western valleys were based on interzonal exchange between independent polities and were not colonialist relationships.

Though ethnohistorians and archaeologists may debate whether the Aymara polities held primarily direct colonial relationships outside of the Lake Titicaca Basin prior to the 15th century, with the arrival of the Incas, Aymara colonial relationships were cemented in Moquegua and other lower altitude regions. I now turn to the ethnohistory of the Lake Titicaca Basin under Inca administration.

THE INCA CONQUEST

In the 15th century, north of Lake Titicaca Basin a group of Quechua speaking peoples known as the Incas was gaining political influence and military power. Society in the Andean highlands, the altiplano, and in the Lake Titicaca Basin during the era of the Aymara polities was a time of wars, or the *auca runa*, the age of warriors (Murra 1986). Several polities were competing against each other in the Peruvian highlands, but by the early 15th century the Incas had emerged as the most powerful among them. By the middle decades of the 15th century, the Incas had expanded into the northern highlands of Peru and were beginning to penetrate south towards the Lake Titicaca Basin (Klein 1992).

Like their northern neighbors, the Aymara Kingdoms were warring polities as well. Murra (1986) notes that the offer of protection and peace between warring polities was an effective part of the Inca campaign. This *Pax Incaica* would have regulated the warring between highlands rivals and redirected violence to the conquering of new territories and the military rule of the peripheries (Murra 1986). By the 1460s the Inca expanded their political influence over the Aymara Kingdoms, which were unable to unite against the Inca. The Aymara had probably posed the strongest military threat to the Inca Empire. However, the political savvy of the Inca rulers played the divided kingdoms against each other until all eventually fell to Inca domination by the end of the century (Klein 1992).

Yet the Aymara were not content to remain as subordinates. Oral tradition records that soon after they were conquered several Aymara lords rebelled while the Inca was having troubles in the eastern Andes (Klein 1992). The Inca ultimately defeated the Aymara rebellion and finished conquering the lesser Aymara polities south of the lake, and the entire Aymara territory was garrisoned by Inca troops. Building on the Andean concept of vertical complementarity, the Inca extended this from strictly agricultural uses for accessing different ecological zones into a method of political control (Murra 1984). Colonists from other parts of the Inca Empire were settled in the Aymara territory, while many Aymara were relocated to the coast and to the jungle to cultivate tropical plants. However, the Aymara colonists remained under the authority of their former chiefs and local ruling dynasties remained intact (Tschopik 1946). The Aymara served the Inca through *mit'a* labor obligations. *Mit'a* labor was a form of tribute paid in actual human labor that was worked in a rotation cycle with other ethnic groups. *Mit'a* obligations for the Aymara included a rotation in the military where they fought under the command of their own leaders and using their own weapons (Murra 1986).

The Inca changed little of the social and political organization of the Aymara Kingdoms. The Aymara retained their traditional rulers and the Inca extracted surpluses through tribute payments and *mit'a* labor obligations. The region of the altiplano and highlands south of Cuzco was organized into its own province known as *Kollasuyo*. However, the Aymara resisted Inca cultural domination and were able to retain their own language, as well as their autonomous social and political structure. Even after the Spanish conquest and subsequent "Qhechuanization" of native populations for administrative purposes, the Aymara retained their language and culture (Klein 1992).

THE SPANISH CONQUEST

Following the fall of Cuzco to the Spaniards, and upon hearing of a temple that was covered in gold on an island to the south, the Spaniards sent reconnaissance expeditions into the altiplano and the Aymara territory of the Lake Titicaca Basin. After the final defeat of the Inca in 1538, the Lupaqa Kingdom declared its independence from the Inca and attacked their age-old enemies, the Colla Aymara. The Colla appealed to the Spaniards for help and Pizarro marched across Bolivia subduing all of the Aymara Kingdoms, eventually conquering the Cochabamba Valley (Tschopik 1946).

For the next decade the southern Andean highlands, which became known as the Charcas region or Upper Peru during the colonial period, was ignored by the Spanish while they fought amongst themselves for control of Lower Peru. However, this changed in 1545 when rich veins of silver were discovered in the mines of Potosí on the southern altiplano. The Lima-based Spanish authorities sent a new expedition into the Charcas region that founded the city of La Paz in the heart of Aymara territory. La Paz quickly became a key city in the transport of goods to and from the mines, and eventually became a commercial center in its own right (Klein 1992).

Though the mines were the primary concern of the earliest Spanish colonists in the Charcas region, they did acknowledge another rich resource available in the area—native labor. The Lake Titicaca Basin and northern altiplano were one of the most densely populated areas in the Andes. The Spaniards attempted to rule this population indirectly by leaving the lands to the local ethnic groups and continuing the Inca pattern of indirect rule. Thus the Aymara *ayllus* remained intact and the local leaders, the *kurakas,* retained their positions of authority. In return, the goods and labor formerly owed to the Inca were redirected into Spanish hands. Most native lands were divided into land grants called *encomiendas* that were governed by an *encomendero* who collected labor taxes and local goods for the state and was entrusted to pay for religious instruction and to acculturate local natives to Spanish rule. In return, the *encomendero* was granted the rights to the labor and local goods of the people who lived on his lands. This created a local Spanish elite who became the governing authority in their local regions and had huge labor reserves at their disposal (Klein 1992).

Murra's (1968) ethnohistoric research on the Lupaqa Kingdom on the eastern shores of Lake Titicaca describes Aymara life during the early colonial period. According to Murra, the Lupaqa polity was divided into two moieties with separate leaders who had descended from a line of leaders from pre-Inca times. Unlike most of their Aymara neighbors, the Lupaqa were not subject to an *encomienda* and paid their tribute directly to the crown. At the time of the *visita*[8] of 1567, the Lupaqa collectively held hundreds of thousands of alpacas and llamas that could be used to store wealth for times of drought, frost, or other calamities. At that time, camelids were also being used to meet Spanish tribute demands, and even after 35 years of Spanish rule, they still had enough animals to sell when annual tribute obligations were severe (Murra 1968)

The Lupaqa polity of 1567 also practiced direct vertical control of lands in distant ecological zones as an archipelago. Besides the nucleus on the altiplano around the city of Chucuito, the polity had a series of far off enclaves on the Pacific coast. The neighboring Aymara polity of Pacajes on the southern shores of Lake Titicaca also had settlements interspersed in the same coastal areas creating a truly multi-ethnic region. These Aymara polities suffered a severe loss when Spanish authorities granted the land and labor in these colonies to Spanish rule as an *encomienda* (Murra 1968).

The mid-1500s were the high point in *encomendero* power on the altiplano with the average *encomienda* in the altiplano region of Upper Peru having over 800 native residents. Though this system of organizing rural life and extracting labor was common Spanish practice, the growing need for labor in the mines of Potosí presented a new problem to the Spaniards and required the creation of a new set of institutions and practices to supply the mines with adequate labor. The system of *corvée* labor designed to supply labor to the mines required the complete reorganization of rural life, local customs, and Spanish administration. The reforms implemented by Francisco de Toledo during his reign as viceroy in Lima attempted to address this problem and proved to be the foundation for centuries of colonial administration (Klein 1992).

TOLEDO REFORMS AND COLONIAL ADMINISTRATION

The Lima-based Viceroy Francisco de Toledo faced an enormous task of restructuring indigenous and colonial society in Upper Peru. From 1572–76, Toledo visited Upper Peru and began the task of reorganizing society in order to meet the tribute and labor demands of both the crown and the local colonials. But Toledo faced two major problems; one was the severe decline of the native population (Cook 1981:94), and the second was the hostility from colonials who were being pressured by the crown to give up their privileged *encomiendas* (Klein 1992).

One of Toledo's solutions was to "reduce" the sparse and scattered native population into permanent settlements and to transform the native *ayllus* into village based communities. The aim of Toledo was to regroup the dispersed populations of highland *ayllus* into large permanent settlements, or *comunidades indígenas*, making them easier to manage and tax. Toledo rapidly created *reducciones*, or new towns, by reducing many scattered settlements into fewer, larger towns (Klein 1992).

On the altiplano newly reduced towns were founded at regular points around Lake Titicaca (Painter 1991), including the colonial town of Tiahuanaco. To govern the newly created towns, Toledo organized a system of indirect rule that granted local autonomy to indigenous communities. Local leaders were elected by the *originarios,* the original community members (Klein 1992). These local administrators were in charge of land divisions and distributions, local justice, the collection of taxes, and the

organizing of labor for the mines. Most communities probably continued
in the pre-Hispanic mode of electing the older, most experienced, and most
successful men as leaders (Klein 1992). On the altiplano, community leaders were called *jilakatas* following the *ayllu* leaders of the previous Aymara
polities.

There continued to exist the local indigenous nobility known as
kurakas who also continued to play a role in rural politics. The *kurakas*
usually oversaw several communities and held their own private estates
with access to labor within the native communities. While *kurakas* were
landowners and exploited local labor resources, they too were also heavily
taxed by the Spanish and were required to guarantee that all local taxes
and labor tribute were paid. The *kurakas* relied on the *jilakatas* to carry out
the Spanish demands for tribute and labor, since their personal estates and
wealth were liable should the communities fail to meet their obligations.
Regional Spanish authorities, known as *corregidores de indios*, were in
charge of collecting taxes and extracting labor at the district level and often
paid for their post by forcing local indigenous subjects to buy imported
goods. The forced sale of goods imported by *corregidores* was an enormous source of wealth for these Spanish officials, creating a lasting hatred
of them by the local populations (Klein 1992).

The highland Aymara struggled to retain control of their lowland
colonies and to preserve their ecologically diverse archipelago. As a result,
many of the new towns created by Toledo were subsequently abandoned,
and many of the valley and lowland communities were never completely
separated from their highland *ayllus* (Klein 1992). The amount of control
exercised by highland Aymara groups in the lake basin over their colonists
in the eastern valleys varied considerably (Saignes 1995). However, all ties
by the highland ethnic groups to the western valleys were apparently sev-
ered when these lands were given to Spaniards in earlier *encomiendas*. In
the end, Toledo's system of concentrated permanent settlements cut off
from direct access to distant ecological zones eventually became dominant
throughout the Andes (Klein 1992).

The other major task of Toledo's reforms was to provide adequate
labor to the mines from the greatly diminished native populations. To do
this, Toledo developed a *corvée* labor system, referred to as the colonial
mita,[9] which extracted forced labor for use in the mines at Potosí. Toledo
based his model on the pre-Hispanic Inca *mit'a* labor tribute, with ethnic
groups from Cuzco to Potosí subject to this forced labor draft, including
the highland Aymara of the Lake Titicaca Basin. In this rotational labor
draft one-seventh of all adult males were subject to a year's service in the
mines, serving no more than once every six years. The mines were obliged

to pay the *mitayos* (*mita* laborers) a small wage, but this amount did not even meet their basic subsistence needs. In reality, the home communities of the *mitayos* paid for the food and other needs of the laborers, as well as helped maintain their households and lands in the community while they were away (Klein 1992). Often whole families migrated during the *mita* labor draft taking with them their livestock and their own food supplies. Since the densely populated Aymara territory of the Lake Titicaca Basin bore a heavy labor obligation, large numbers of people were forcefully recruited and relocated to the southern altiplano mines (Painter 1991).

With Spanish conquest also came the dramatic decline of the native population (Cook 1981). While the total number of tribute paying *originarios* continued to decline throughout the 16th century and into the 17th century, the amount of tribute required of them did not subside making the burden on the remaining population intense by the end of the 16th century (Klein 1992). This growing pressure on the *originarios* of the indigenous communities caused changes in the social and economic bases within the communities. It was the *originario* class that was responsible for paying all tribute taxes and was subject to the *mita* labor draft. Since native populations might exempt themselves from tribute obligations by establishing new residences elsewhere, there quickly developed a migrant population (Klein 1992; Painter 1991). These *forasteros* (literally foreigners) were given access to smaller plots of land or simply took up resident as landless laborers in indigenous communities (Klein 1992).

The same demographic and economic pressures that created the *forastero* class, also created a new group of natives not tied to any indigenous community, but instead lived on Spanish estates. Known as *yanaconas*, by the late 16th century this term simply meant landless workers (often *ex-originarios*) who were willing to work on Spanish estates in exchange for usufruct rights to land (Klein 1992).

By the mid-17th century there was a demographic shift of population from the highlands to the cities and lowland valleys (Saignes 1995). In the regions that provided *mita* laborers, such as the Lake Titicaca Basin, the drop in population was dramatic. The Lupaqa and Pacajes ethnic groups lost between three-quarters to four-fifths of their tributary populations and gained only a small amount in *forasteros* and *yanaconas*. This decline was not simply a result of disease and exploitation, but also due to out-migration. For example, by 1658 the town of Tiahuanaco had only 9 tributaries, while in the eastern valley provinces of Omasuyos and Sica Sica the populations of *forasteros* and *yanaconas* swelled, probably due to highland in-migration (Saignes 1995).

During the colonial period, the rural areas of Upper Peru contained over 90 percent of the population, of whom 90 percent were monolingual speakers of indigenous languages. In the Lake Titicaca Basin, the Aymara language continued to dominate rural life, despite further attempts at Quechuanization by the Spanish who continued to emphasize the Quechua language as the *lingua franca* across diverse ethnic groups. By the 1580s, the continued persistence of Aymara finally resulted in the publication of the first Catholic catechism in that language in 1584. By the early 1600s the Jesuits published complete grammars and dictionaries of the language (Klein 1992). Aymara culture and language continued to serve as an ethnic marker and as a symbol of resistance on the Aymara altiplano during the frequent outbreaks of violence throughout the colonial period.

LATE COLONIAL SOCIETY AND "THE AGE OF ANDEAN INSURRECTION"[10]

While booming silver production during the 16th century fueled the demand for mine labor, by the mid 17th century production had peaked and the Upper Peruvian economy began to contract dramatically and would continue to decline for over a century. A direct result of this economic decline was the depopulation of cities in Upper Peru, particularly the mining centers of Potosí and Oruro. The city of La Paz on the northern altiplano was the sole exception. La Paz continued to grow throughout this period, so that by the end of the 18th century it had surpassed Potosí to become the largest city in Upper Peru and the hub of commerce and administration. This modest growth during a period of general economic decline is attributed to Aymara production on the altiplano, as well as the thriving agricultural production in the eastern valleys (Klein 1992).

The general economic decline of the 18th century coincided with what Stern (1987a) calls the "Age of Andean Insurrection," a decades long period of rebellious ferment leading up to the full-scale civil war of 1780–82 that challenged the existing colonial order. The insurrectionaries were primarily Quechua and Aymara speaking peoples whose lives were directly influenced by the shifting political economic contexts. Historians continue to debate the causes of this massive pan-Andean revolutionary movement. Factors include the growth of native populations and the reorganization of indigenous communities due to increases in the *forastero* class (Klein 1992, 1993), and the reforms of the taxation system that led to increased hostilities between *corregidores* and natives (Stern 1987a).

Ethnohistorians have also begun to focus on the specific cultural contexts in the years leading up to and including the civil wars of 1780–82 (Campbell 1987; Robbins 1994; Stern 1987a; Thomson 1996; Valle de

Siles 1990). One reoccurring theme in this age of unrest is the appearance of a pervasive pan-Andean messianic myth the predicted a utopian future based on the reversal of the colonial social order under the leadership of a reincarnated Inca-King (Campbell 1987; Stern 1987), the myth on *Inkarrí*. The *Inkarrí* myth envisioned the return of the ancient Andean creator who would restore the Inca reign by reincarnating himself as a direct descent of the Inca royal lineage.

    This myth spurred the actions of several revolutionary leaders who took the name Inca or in some way signified their link to the Andean Creator as the true king returned to earth. The myth of a returning Inca-King had the symbolic power to politically unite diverse Andean ethnic groups, although it could not overcome all inter-ethnic rivalries. However, Campbell (1987) argues that the symbolism of the *Inkarrí* myth was challenged in the Aymara heartland, where claims to rule by a Cuzco-based Quechua leader were not supported. The Aymara version of this messianic myth placed the returned kingdom in the hands of an Aymara leader. This myth was first realized when Tomás Katari, an Aymara *cacique* from Chayanta, led a rebellion in Upper Peru in 1777. Following the rebellion of Tómas Katari, the Quechua leader Túpac Amaru led a rebellion of 1780 in the central Peruvian highlands. Under the auspices of the *Inkarrí* myth, both of these leaders challenged local authority on the grounds that they were immoral and unjust (Szeminski 1987:174).

    The dramatic events of 1780–82 demonstrate the power of the *Inkarrí* myth to rally popular support that translated into a pan-Andean revolution targeting Spanish authority in the Peruvian highlands and throughout Upper Peru (Szeminski 1987). Though the pan-Andean rebellion of 1780–82 across highland Peru and Upper Peru failed to unseat Spanish authority, Stern maintains (1987a) that these rebellions harbored the seeds of the coming national revolutions for independence. The *Inkarrí* myth was a "protonational" symbol that linked native Andean political participation to a new era of Andean social order. Stern (1987a) contends that Andean peasants in the colonial period were political actors who aspired to be part of a wider culture and that indigenous peoples were not merely local reactors unable or unwilling to envision themselves as part of a nation or a state. In fact, the *Inkarrí* myth itself demonstrates an idyllic Andean rendering of the new pan-Andean state that was to come.

    In the Aymara heartland of Upper Peru, the *Inkarrí* myth was ultimately manifested in the rise of Túpac Katari, a commoner by the name of Julian Apaza who combined the names of his predecessors and claimed to be the Aymara leader Tomás Katari reincarnate. Under the leadership of Túpac Katari, the Aymara of the Lake Titicaca region and the eastern yungas region rebelled and laid siege to the city of La Paz (Klein 1992). The city of La Paz, located in a deep canyon cut into the altiplano floor, was laid siege by hundreds of thousands of Aymara and Quechua peoples. The rebels camped on the high plains that surrounded the city, blocking all access routes in and out of the canyon. Not until November of 1781 when Túpac Katari was taken prisoner and executed, did the rebellion begin to disintegrate (Valle de Siles 1990).

The *Inkarrí* myth and the rebellions of 1780–82 are key moments in Aymara moral memory. The rise of Túpac Katari as a leader of the Aymara, separate from the Cuzco-based Quechua speaking Inca leaders, signals the reawakening of nativist Aymara political consciousness. By supporting the Aymara leader Túpac Katari, and not the Quechua Inca Túpac Amaru, the Aymara were fighting to establish a new utopian state free of Spanish and Inca political control. This renewed political consciousness paved the way for the independence movement that was to shake Bolivia and the rest of Latin America in the first half of the 19th century. Contemporary Aymara indigenous social movements continue to invoke the image of Túpac Katari in their struggles for political power on the altiplano and in the city of La Paz.

Along with a renewed sense of Aymara nativist political conscious-ness, Thomson (1996) argues that there was a fundamental shift in com-munity power relations in the Aymara heartland of the altiplano following the rebellion. Rural indigenous power had previously been held by noble-born Aymara *caciques,* whose authority and power was upheld by the Spanish state. The *caciques* were private landholders who exploited native labor and used various forms of coercion to meet Spanish labor and trib-ute demands. Following the failed rebellions, Thomson (1996) argues that community representatives began to challenge the traditional authority of the *caciques* and political authority was transferred to the local communi-ties and the community elected *jilakatas.* The transfer of power from the noble-born *caciques* to the local communities included the taking over of community tribute collection in some cases, representation of local disputes and legal denunciations with the state, as well as the governing and admin-istration of the indigenous communities. How these locally elected *jilakatas* handled the trials of a new republican state government, a national depres-sion through much of the 19th century, and the growing loss of indigenous lands varied from community to community but general trends will be examined in the following section.

THE 19th CENTURY REPUBLIC OF BOLIVIA

The beginning of the 19th century was a period of continued economic depression made worse by the damaging effects of the wars for independence from 1810 through 1825, after which Bolivia would become a new republic. Silver production during the late 1790s decreased dramatically due to disruptions caused by the Napoleonic wars in Europe, and a severe depression in the international markets effectively cut off capital to the costly Upper Peruvian mines causing a further drop in production. By the early 19th century, silver production was in deep decline and Upper Peru was hit by a series of harvest failures and epidemics (Klein 1992).

Fighting broke out in South America in La Paz in 1809, followed by the first declaration of independence from Spain by *Paceño*[11] rebels. Though quickly suppressed, this initial declaration began the long period of war with Spain for independence in her American colonies (Klein 1992). Throughout the fighting, many cities in Upper Peru were sacked repeatedly by both Argentine rebels and Spanish forces. The countryside was also plagued with violence by local rebellions resulting in haciendas being razed, isolated mines destroyed, and leaving the rural economy in turmoil. The fighting finally ended in Upper Peru in 1825 when Spanish armies were defeated and royalist leaders in La Paz surrendered to the rebels (Klein 1992).

When the fighting finally ceased, the new Republic of Bolivia was facing severe economic hardships. The mining industry continued in a serious state of depression throughout the first half of the 19th century. Bolivian urban centers all experienced significant population loss, with the exception of La Paz, and the new republic was more rural and subsistence oriented than anytime in its Spanish colonial past. During the period of 1830–1860, the decline in the mining sector of the economy reduced the economic pressure on the free Indian communities, resulting in an overall increase in their level of income. Thus the indigenous communities were able to increase their internal trade due to their higher relative income levels, and fully supported the regional economy. This regional market led by the highland Aymara fueled the modest growth of the city of La Paz during a period of general urban decline (Klein 1992). The continuation of the colonial tax singled out indigenous communities and hacienda workers, though it left the corporate communities in tact and abandoned the liberal model of private property. By the 1860s, the importance of the tribute tax began to decline as the mining sector started its slow recovery and the government began to collect more revenue from mining and other commercial activities. Thus Klein (1993) argues that the final acceptance of the liberal model of private property and the onslaught of hacienda expansion in the later decades of the 1800s was directly tied to the national economy.

Recent works on Bolivia (Calderón 1991; Grieshaber 1980; Jackson 1989; Langer 1989; Larson 1998; Platt 1982) and Peru (Jacobsen 1993; Mallon 1983; Nugent 1997) have sought to decipher the ways that indigenous communities resisted hacienda expansion, and to assess the level of political consciousness of indigenous inhabitants in the 19th century. On the altiplano, the Aymara communities thrived despite a government that envisioned private property as the foundation of its liberal land policies. For example, Grieshaber (1980) demonstrates that throughout the mid-19th century the Aymara communities on the altiplano controlled more rural labor than the haciendas.

However, by the 1880s the ability of the indigenous communities to resist the encroachment of haciendas began to weaken in the face of renewed assaults on their land base, particularly in the Aymara altiplano of La Paz (Rivera Cusicanqui 1987). While the indigenous communities still held half the land and rural population in 1880, by 1930 they had been reduced to less than a third (Klein 1992). The result of this land loss was increased migration to the cities and the expansion of the urban populations. The expansion of the haciendas also prompted the rise of the *colonato* class. The *colonato* is a system of tenant farming, with access to land granted in return for work on the hacienda, similar to the colonial *yanaconas*. Throughout this period the number of *colonos* continued to increase, with their class interests at times conflicting with that of the indigenous communities (Rivera Cusicanqui 1987). The exception to this trend of hacienda expansion is in the province of Cochabamba, where haciendas declined and smallholder agriculture thrived (Jackson 1989; Larson 1998; Sanabria 1993).

Tristan Platt (1993, 1987, 1982) has explored the political roles of the Aymara in the southern altiplano province of Potosí during the 19th century. Platt argues that local Aymara, who had participated in the various rural rebellions of independence, quickly became disillusioned of the "social justness" of the new republic. Platt (1987) goes into detail outlining peasant political participation throughout the 19th century, culminating in the indigenous participation in the civil war of 1899. Ultimately, Pablo Zárate Willka, an Aymara *cacique* from La Paz, led an uprising on the side of the Liberals during early phases of the war and was the symbolic leader of the Aymara. However, the Aymara rebels soon developed their own agenda that included the restitution of lands, resistance to hacienda growth, and the establishment of indigenous autonomy. Obviously, these objectives clashed with the Liberal plan and brought renewed attention to the ethnic tensions that had been building as a result of the changing economics, particularly on the densely populated altiplano of La Paz (Rivera Cusicanqui 1987). Following the Liberal victory, the indigenous troops who fought on the Liberal front were quickly disarmed and their leaders executed. The expansion of haciendas continued under the Liberal government in the first decades of the 20th century and led to increasing conflicts with the indigenous communities (Klein 1992).

THE EARLY 20th CENTURY AND NATIONAL REVOLUTION

At the turn of the century, the Liberal Party took control of the government though life did not change much for indigenous inhabitants. The Liberals continued to support the confiscation of indigenous land and the number and size of haciendas grew unabated. For example, in the Pacajes province of La Paz approximately 44,687 hectares of land were sold from 1901 to 1920, compared to the 33,401 hectares sold from 1880 to 1900 (Rivera Cusicanqui 1978:106). The national economy also continued to grow, though the once dominant silver mines had been replaced by the booming tin mining industry. With a revived economy, the Liberal regime committed itself to public works, particularly the construction of railroads. The new rail system created links between the mining centers, the regional cities of Cochabamba and Sucre, and the Peruvian rail network via the Guagui–La Paz rail line that transected the northern altiplano, through the town of Tiahuanaco (Klein 1992).

After the military defeat of the Zárate Willka movement, the Aymara resistance movements were again fragmented. However, beginning in 1910 Indian rebellions once again began to break out among the Aymara in the altiplano provinces. For example, there was a rebellion in Pacajes in 1914, a revolt of tenant farmers in Caquiaviri in 1918, and an uprising and massacre of community Indians in Jesus de Machaca in 1921. Though these and other rebellions did not represent a clearly united Aymara struggle, the broad range and frequency of outbreaks and uniformity of demands does indicate a growing cohesion of interests. Silvia Rivera Cusicanqui (1987) refers to these violent outbreaks from 1910–1930 as a "cycle of uprising" that had common ideological, political, and organizational characteristics. The collective Aymara struggles on the altiplano were not just a response to hacienda usurpation of Indian lands. There was also a basic conflict between indigenous agricultural producers and the monopoly that the rural townspeople had over the regional trade networks (Rivera Cusicanqui 1987).

In the 1920s political leadership again returned to civilian Republican governments and political ferment was heating up in other arenas besides the rural unrest of the indigenous communities. Marxist and other socialist political philosophies were finding audiences with intellectuals and the urban middle classes, and new parties began to form that challenged the established elitist Liberal and Republican two-party system. In 1921 a National Socialist Party was formed with some support from labor that raised the issues of servitude on the haciendas (*pongueaje*) and the legal rights of the indigenous communities. In 1928, a group of radical students formed the first National Federation of University Students of Bolivia (FUB). Though small in number, these two groups were the foundation for the rise of revolutionary political parties following the Chaco War and for the first time there were now persistent calls for agrarian reform and the end of colonial rural servitude (Klein 1992).

When a border dispute with Paraguay escalated into an act of war, Bolivia was plunged into her costliest and most debilitating war in her young history. The widely held belief that the conflict was over international oil interests demoralized the Bolivian army, and shaped the formation of the radical political leanings of the "Chaco generation" that came to maturity in the years following the end of the war in 1935. Radical political ideas engaged the disillusioned and bitter war veterans, which had previously only been voiced by young radicals and intellectuals. Following the war, Bolivia was mired in another deep depression brought on by the global depression, which finally brought an end to hacienda growth. By this time, the number of landless *colonos* far outnumbered the landowning indigenous community members. This landlessness provided a continuing source of conflict following the Chaco War, when returning soldiers faced the double obstacle of landlessness and decreased opportunities for urban employment due to the national depression (Klein 1992).

In the 1930s new political parties began to form based on diverse and radical political ideologies. The two most important among them were the Partido de la Izquierda Revolucionaria (Revolutionary Leftist Party [PIR]) and the Movimiento Nacionalista Revolucionario (Nationalist Revolutionary Movement [MNR]), the latter of which drew support from the middle classes, the generation of post-Chaco veterans, as well as the emerging workers' and peasants' movements (Rivera Cusicanqui 1987).

Newly created peasant labor unions set about the task of organizing the peasants of the altiplano and eastern valleys (Dandler and Torrico 1987; Rivera Cusicanqui 1987). Some of the primary organizers of the unions on the altiplano were the Aymara *caciques*, perhaps renewing their role as revolutionary leaders as they had been in previous ethnic struggles. For example, Marka T'ula, an Aymara *cacique* from La Paz, organized strikes on several highland haciendas, while maintaining contacts with urban workers and students. Having been active in protests before the war, T'ula eventually became a leader in the Federation of Unions of Oruro (Rivera Cusicanqui 1987)

In 1945 the 1ˢᵗ National Indigenous Congress was held in La Paz during the MNR coalition government of Villarroel. With state support, peasant leaders arrived in La Paz and were admitted into Plaza Murillo, the seat of state government, for the very first time. When Villarroel was later dragged by *Paceño* mobs and hung by a lamppost in that same plaza in 1946, he became a martyr to revolutionaries (Klein 1992:221) and found a place in the symbolic moral memory of the rural peasantry alongside the memories of Túpac Katari and Zárate Willka. For *Paceño* inhabitants, Villarroel had touched their own deep-seated fear of the peasant masses and revisited the memory of the siege of Túpac Katari in 1781.

Through a series of uprisings across the altiplano in Aygachi, Pucarani, and Los Andes (Rivera Cusicanqui 1987) and in Cochabamba (Dandler and Torrico 1987), the rural peasants retaliated after the downfall of Villarroel and the dismantling of his reforms to the *colonato* system. By 1950 MNR had reemerged as a political threat to the conservative regimes, by drawing on the support of virtually all organized labor as well as much of the middle class. In 1951 MNR ran Víctor Paz Estenssoro as their candidate for president and won the election with a majority of votes. When the army intervened to prevent his ascent to power, MNR resorted to armed revolt to uphold the election results and after three days of fighting, much destruction, and the loss of over 600 lives, MNR finally came to power in April of 1952 (Klein 1992)

With the end of the revolution and the resulting agrarian reform of 1953, Bolivia entered into a new political era. The indigenous population was finally granted suffrage and began to have their political and economic claims heard. Education was nominally extended to all citizens throughout the countryside and a new class of smallholder agriculturalists sprang up almost over night. The Aymara of the Lake Titicaca Basin had completed a remarkable millennium of transitions from being at the center of a civilization based on intensive agriculture, to independent polities that drew more on pastoral resources, to an exploited ethnic group under foreign colonization, and continued class based exploitation after national independence. In 1952, the rural peasants were granted the opportunity to participate in the national government, though this transition to democracy and the political representation of the peasantry is still a work in progress.

NOTES

[1] *Mestizaje* is a term that refers to political discourses of ethnicity in Latin America that attempt to co-opt and integrate both Spanish and indigenous identities.

[2] Note that the image of Tiwanaku as a state level society, though contested by some archaeologists, is the dominant image, that is being reproduced and supported by the Bolivia state.

[3] Again, note that the director of the NGO was also the director of the National Institute of Archaeology.

[4] *Indigenismo* was a literary and political discourse concerning ethnicity and race that was prominent in the early decades of the 20th century in Latin America.

[5] I use the term "moral memory" following Stern (1987a) to discuss the interplay between history and group consciousness with emphasis on the sense of historical and moral outrage in response to material exploitation. Similar terms include "cultural memory" or "social memory" (Abercrombie 1997:21), however these terms do not explicitly recognize the role of social power in the creation and control history, myth, and memory.

[6] The *altiplano* refers to the high plateau of the central Andes, which includes the Lake Titicaca Basin.

[7] *Kuraka* is the Quechua name for these indigenous lords. There were also called *mallku* in Aymara and *cacique* in Spanish. These three terms can be used interchangeably.

[8] *Visitas* were administrative reports compiled in the field by Spanish officials in the early decades after the Spanish conquest (Murra 1968).

[9] I use the spelling *mit'a* to refer to the Inca labor tribute and *mita* to refer to the Spanish labor draft.

[10] Stern (1987a:35); Thomson (1996:12) calls this period the "Age of Insurgency" emphasizing an ongoing struggle of power relations between caciques and local indigenous communities.

[11] *Paceños* are people from the city of La Paz.

CHAPTER THREE

The Contexts of Agricultural Development: Agrarian Policies, Indigenous Social Movements, and Sustainable Development

D EVELOPMENT AND DEVELOPMENT PROJECTS ARE NEVER DEVOID OF
politics, economics, or cultural factors. Development projects in
third world countries have always been tied to foreign political
interests and economic policy, as well as shaped by both international and local economic trends, and cultural contexts. I argue that the
rehabilitation of raised field agriculture in Bolivia was not simply a spon-
taneous or well-timed discovery by social scientists and researchers, and in
fact the raised fields had been known about for at least 20 years (Smith et
al. 1968). Further, the enthusiastic reception of raised fields as a development project in the late 1980s was not merely a result of a timely "new discovery" that only needed to be implemented in order for it to succeed.

By representing raised fields as a "lost technology," researchers
made several assumptions about the local residents of the Bolivian altiplano who are the alleged heirs to this ancient agricultural system. One
implication is that the primary reason raised fields were not in modern
practice was because local Aymara inhabitants simply no longer knew how
to build or cultivate the fields as their ancestors had before them. Thus
implying that researchers merely needed to reintroduce this "lost" and
"indigenous" agricultural method, and re-educate the locals on its usage,
so that they could once again reap the benefits of this successful pre-
Hispanic agriculture system.

Through depicting the raised fields as a lost technology and somehow natural or ancestral to the Lake Titicaca Basin, researchers and development personnel have either overlooked or downplayed the monumental
changes that have taken place within economic and political contexts since

the decline of the Tiwanaku civilization. Foremost among these changes was a shift from a non-monetary economic system of vertical control with tribute based on labor extraction, to a market driven nation-state based on monetary taxation of smallholder farmers. A second key change has been the long-term affects of political oppression by Spanish and mestizo elites on indigenous peoples, who have only recently begun to dismantle and disarm the hegemonic nationalist discourse through an "ethnic awakening" (Albó 1995:43).

I contend that raised fields were not simply a lost technology that once "rediscovered" merely had to be rehabilitated in order to be successful. In fact, I will argue that the success of raised fields in generating intense enthusiasm with funding agencies and development NGOs was based on several developing trends that were on the rise in the late 1980s in Bolivia and globally. Raised field agriculture first gained the attention of researchers who hypothesized that the fields were the economic foundation for the advancement and florescence of the once mighty Tiwanaku civilization. This hypothesis quickly came to be regarded as fact by development groups, who became interested in the raised fields and the development project that later ensued.

However, by examining the political, economic, and social trends in Bolivia and globally, I will connect the raised field rehabilitation project with broader agendas, such as government economic interests and development policy, indigenous social movements and ethnic politics, and the rise of development theory predicated on the concept of "sustainability." The raised field rehabilitation project had several discernible characteristics, such as its low economic and technical inputs, its indigenous origin and "ecological sustainability," and its imagined links to a primordial past that fit with the expectations and agenda of academic researchers and development personnel. It was this combination of interconnected trends that heralded in the raised field rehabilitation project in the late 1980s.

In this chapter, I examine the contexts of agrarian reform on the northern altiplano and in the community of Wankollo by investigating local land distribution and the politics of becoming smallholder landowners. Second, I trace broader political and economic trends in Bolivia following agrarian reform and through the neo-liberal economic policies of the mid-1980s. Third, I explore the pervasiveness and application of ethnic claims in an indigenous political movement in La Paz. Fourth, I critically review the history of development and the turn towards environmentally friendly projects via "sustainable development." The concluding section summarizes the ways that these developing trends created the dramatic setting for the raised field rehabilitation project to enter the Bolivian development scene, and how it set predetermined characteristics of what a successful, desirable, and "appropriate" development project would be for the Bolivian altiplano.

THE AGRARIAN REFORM IN WANKOLLO AND THE LAKE TITICACA BASIN

At the time of the April Revolution of 1952 led by the *Movimiento Nacionalista Revolucionaria* (Nationalist Revolutionary Movement or MNR), Wankollo was a hacienda in the Tiahuanaco Valley that lie on the southeast border of the town of Tiahuanaco. The hacienda was less than 70 km from La Paz and was linked to the city by a rail line that ran from the port of Guaqui on Lake Titicaca across the northern section of the hacienda. The hacienda was also connected to the capital city by a road, which served as the boundary between the hacienda and the two northern neighboring haciendas. Though rural, the hacienda of Wankollo was certainly not isolated or remote since it had direct access to the city and to the international border with Peru.

Prior to the agrarian reform, the Aymara speaking residents who lived on the hacienda at Wankollo were granted access to plots of land in return for work in the landowner's fields and in his home. While the Catari Valley to the north had had haciendas since the colonial period, most of the haciendas of the Tiahuanaco Valley and Desaguadero Valley were from the post-Independence era of hacienda expansion. These former *ayllus* of the Tiahuanaco and Desaguadero Valleys had been traditional strongholds of the lake basin Aymara, particularly during the Colonial period and the 19th century (Rivera Cusicanqui 1987).

The Bolivian agrarian reform that was enacted in August of 1953 finally put an end to the hacienda system on the altiplano, thus legitimizing the take over of the haciendas by their peasant workforce. Almost overnight, the Bolivian altiplano went from a classic *latifundia* system of large haciendas into a *minifundia* system of many thousands of altiplano smallholders who were now called "*campesinos,*" or peasants, rather than "*indios*" (Carter 1964; Thorn 1971).

Under the agrarian reform law, those few remaining free indigenous communities that had managed to withstand hacienda encroachment were made owners of their collective lands. The agrarian reform also attempted to reinstate lands usurped from them as well, though not with much actual success. The apportionment and distribution of the community lands among the members was legislated to follow "custom" giving free reign to the communities to continue their previous land tenure arrangements. All indigenous communities were supposed to establish some common pastures for community use if there was no land already in communal pasturage (Carter 1964). In the Bolivian lake basin, the land that was held in common solely for pasturing animals typically was the low, marshy areas unsuitable for modern agriculture, where the remains of pre-Hispanic raised fields are often found.

On the altiplano, the redistribution of ex-hacienda lands tended to follow former usufruct rights of residents on the hacienda. In Aymara free communities, the *sayaña* is a parcel of land where a house compound is located and is reserved for use only by that household. An *aynoka*, on the other hand, is a communal holding with access to land either communal or granted on an individual basis. Following the agrarian reform, families that had had *sayañas* on the ex-haciendas usually received titles to these lands and little more. Further, the agrarian reform did little to rectify the inequality of landholdings and wealth differentials between "rich" and "poor" peasants. On the haciendas, the *colonos* had been ranked in regards to access to land, so that not all had equal holdings or equal responsibilities in terms of labor owed to the landowner. In actuality, the agrarian reform served to reinforce these disparities since better off peasants had more access to legal aid for securing titles to land and staging legal battles (Carter 1971:248; Mendelberg 1985). Also, families with large household *sayañas* probably did not lose any land, since *sayañas* were rarely broken down into smaller plots in order to equal-out the holdings of the ex-hacienda workers, even when the plots exceeded the maximum land allotment for the lake region (Carter 1964:76–77). In Wankollo, there was certain inequality of wealth and land distribution, as I will illustrate in the following section.

In many cases the redistribution of the hacienda lands did not come without significant conflict from within Indian society (Buechler 1969:189). In 1953, the hacienda of Wankollo was comprised of 1,930 hectares of land and had a resident population of *"colonos"* (landless workers), who claimed the land of the former hacienda. Following the agrarian reform, a group of landless workers from La Paz, some of whom had served in the Chaco War, contested the claims of the former *colonos* of Wankollo on the assertion that they had been wrongly dispossessed of their holdings by the former landowner. Calling themselves *"ex-comunarios"* of the original Aymara *ayllu,* and claiming that they had been unfairly evicted from their land prior to the agrarian reform, this group successfully lobbied to receive land in Wankollo. As a result, this group of 54 *ex-comunarios* from La Paz received just under 733 hectares of land, compared to the slightly higher amount of 794 hectares granted to the 30 *ex-colonos* of the former hacienda.

However, though the *ex-comunario* group was able to obtain lands in Wankollo, they by no means gained equal sized holdings. Nor was the distribution of these lands based on customs of usufruct rights. Instead, the community was split in half with the *ex-colonos* to the east and the *ex-comunarios* getting the plots on the western half of the community. This struggle between *ex-colonos* and *ex-comunarios* was a fairly common event in some communities in the years following agrarian reform (Carter 1964), though Buechler (1969:191) maintains that the considerable success of the Wankollo *ex-comunarios* in reclaiming hacienda lands was atypical.

The amount and distribution of land granted to the Bolivian peasantry varied based on how the hacienda was categorized in the agrarian reform. For example, in cases where the hacienda was categorized to have some modernization improvements, the landowner was legally allowed to retain a portion of his land. In these cases, the *colonos* of the former hacienda often did not receive any increase in access to land, since they generally only received title to the land which they had previously held usufruct rights. However, on haciendas that were turned over completely to their former workers, most *colonos* experienced significant augmentations to their holdings. Of course, in the indigenous communities, access to land was not significantly altered, except in the few instances when they were able to reclaim land from the haciendas (Buechler 1969; Carter 1964). In Wankollo, all of the land was divided between the *colonos,* the groups calling themselves *ex-comunarios,* and newly entitled persons from both groups, with an additional large section (145 hectares) held in common (called an *aynoka*).

The agrarian reform not only distributed land back to the former *colonos* (and, in the case of Wankollo, to earlier claimants), but also put an end to all tribute and *pongueaje* service.[1] In some respects, this was even more significant than the granting of land ownership and increases in size of holdings to the *campesinos* (peasantry). On many ex-haciendas, and certainly in the few remaining free communities, the amount of land that the *campesinos* had access to did not increase very much. However, with the demise of *pongueaje* service, the ex-hacienda workers had very significant increases in the amount of labor available to households. For example, prior to reform *colonos* in the Lake Titicaca region typically devoted three days each week to the lands and properties of the landowner and during the harvest they worked as long as needed to complete the harvest. Workers also provided their own tools, seed, and oxen, and assumed various other minor tasks such as making *chuño* (freeze-dried potatoes) and working in the manor and town houses of the landowner (Burke 1971:328).

There continues to be a debate about the impact of agrarian reform on agricultural production in Bolivia (Burke 1971; Klein 1992; Mendelberg

1985; Thorn 1971). However, it is difficult to generalize about land and labor productivity as rates varied widely according to factors such as ecological zone, prior forms of tenancy, and population density. I focus specifically on the peasantry of the Lake Titicaca Basin, who generally had access to numerous scattered small plots following the agrarian reform.

In the Bolivian lake basin, the agrarian reform did a rather thorough job of distributing land back to peasant cultivators (Buechler 1969; Carter 1964). However, it was the *hacendados* (hacienda owners) who had controlled the marketing of goods and had been the primary suppliers of foodstuffs to the cities. With the *hacendados* gone, the marketing and distribution systems fell into temporary disarray disrupting the flow of agricultural goods from the countryside (Carter 1971:249; Thorn 1971). A common assumption in Bolivia was that agricultural production had decreased and de-intensified without the *hacendados* overseeing the work of the peasantry (Carter 1964).

However, Burke (1971) maintains that in the Lake Titicaca Basin land actually began to be cultivated more intensively after the reform, as populations on the ex-haciendas swelled to twice their previous numbers with many peasants returning to communities to place land claims. Wankollo is a good example of this, where 54 *ex-comunarios* and an additional 22 new heads of households, returned to the community with their families to lay claim to land. Burke (1971), in his work on four Bolivian ex-haciendas, argues that not only was land worked more intensively following the reform but that increasing amounts of marginal lands also began to be cultivated. Burke concludes that agricultural output actually increased in the Lake Titicaca region in the ten years following the reform. He argues that the reason less produce reached the cities was because the considerably larger rural population consumed most of their produce rather than directing it to the markets (Burke 1971:317).

Recent work measuring sedimentation rates in the southern lake basin (Binford et al. 1996) also suggests that agricultural activities actually increased in intensity in the years immediately following the agrarian reform. A measured rise in the rate of sediments on the lake bottom beginning at about the time of the agrarian reform may be interpreted as an increase in agricultural activities, since the plowing and cultivating of land releases soil into the lake due to erosion (Binford et al. 1996). Thus the increasing rate of sedimentation gives further indirect evidence that agricultural activities and agricultural intensification were on the rise in the lake basin (Swartley n.d.).

In parallel with this agricultural intensification was a marked increase in off-farm employment following the agrarian reform as *campesinos* began seeking seasonal employment in the eastern valleys as agricultural laborers, and in La Paz as seasonal construction workers (Burke 1971; Mendelberg 1985; Wessel 1966). Therefore, much of the labor freed by the removal of *pongueaje* servitude was being redirected into economic activities that increased the wealth of *campesino* families.

In addition to the economic changes taking place in the lake basin after agrarian reform, there were also political changes sweeping the countryside as *campesinos* began the process of organizing themselves into *sindicatos* (community labor unions). The *sindicatos* soon became, and continue to be today, the foundation of rural peasant political organization. Following reform the MNR government, with the leadership of the newly formed *Central Obrera Boliviana* (Bolivian Workers Central or COB), helped to organize the rural peasant unions. Though proclaiming itself politically neutral, in fact the COB was initially a powerful ally of the MNR government and in this fashion the MNR government hoped to maintain control of the peasant unions (Klein 1992, Van Cott 2000a).

POLITICAL AND ECONOMIC CURRENTS AFTER AGRARIAN REFORM

After the agrarian reform of 1953, the *campesinos* temporarily became a relatively conservative political force, creating a stabilizing popular base for the post-revolutionary MNR and the Barrientos regimes. Having attained suffrage and ownership of their land, many rural regions began to focus more on local and regional political rivalries in their attempts to organize themselves. The community level politics involved in allotting land grants to individuals often consumed much of the *campesinos'* attentions, such as the case in Wankollo where final land divisions where not drawn up until 1960. The MNR government realized the importance of this stable and conservative base of support, particularly in the face of losses to its traditional middle class and leftist mineworker support bases (Klein 1992).

When the MNR regime fell apart and was replaced by the populist military dictator René Barrientos in 1964, the incoming General reconfirmed the government's support of the peasantry and the agrarian reform. There continued to be an alliance between the military governments and the peasantry throughout the dictatorship of Barrientos and his successors (1964–1971). This alliance was based on a form of clientalism, which granted government support of the agrarian reform and distribution of government services, such as education and rural union support. In return, the peasantry remained a loyal and relatively stable support to the government, though this support would begin to crumble under the radical leftist military government of Juan José Torres (Klein 1992).

Bolivia's long-term economic problems continued into the 1980s, including her historic over-dependency on a single boom or bust mining industry. The Bolivian economy has long depended on the mining industry as the foundation of its exports. However, as a landlocked and mountainous country, Bolivia suffers from high transport costs for all of her exports. With declining tin quality and lack of modernization, combined with high transport costs, Bolivia's mining industry began to be less and less competitive in the world market. The only alternative export in the 1980s was petroleum (not including illegal coca), which overcame high transport costs since it was piped directly into Argentina (Morales and Sachs 1989).

The economic crisis of the 1980s had its roots in the Bánzer era, a result of a number of fiscal policies, some irresponsible and others merely unsuccessful. During the Bánzer years (1971–78) the economy appeared to be quite good as the GDP (gross domestic product) rose an average of 5.4 percent per year and exports soared to new highs. Yet the gap between government expenditures and government revenues widened as the Bánzer regime continued to spend (Morales and Sachs 1989). Clientalism and political patronage also grew unabated during Bánzer's regime, one indicator being the rapid increase in government jobs from 66,000 to 141,000 between 1970 and 1974 (Jameson 1989:88).

During the Bánzer era, economic development and development projects also began to be shifted away from the altiplano and into the tropical lowlands, as the region began to benefit from Bánzer's political patronage. Though the agrarian reform had brought forward the agenda of eastern colonization and development, under Bánzer the disparity of agricultural credit and development channeled into the altiplano vs. the eastern lowlands became particularly apparent. By 1970–75, the eastern lowlands were receiving nearly 89% of all agricultural credit and 73% of funds from the regional development budget in 1975. This compares to the altiplano area, which received fewer than 5% of agricultural credit from the *Banco Agrícola* and just under 15% of funds from the regional development budget (Dunkerley 1984:221).

Since Bánzer was unwilling to raise the tax base, where government revenues fell short of spending foreign bankers easily stepped in to supply the government with the needed cash (Morales and Sachs 1989). As a result, foreign debt in Bolivia also soared to new highs increasing fourfold during Bánzer's term from $782 million in 1971 to $3.5 billion by 1979 (Jameson 1989:88). By the end of Bánzer's term in office, Bolivia was facing a severe recession and the following years of political instability did nothing do rectify these serious economic problems.

Bánzer was finally forced to step down from government in 1978 and Bolivia began the struggle towards democratic government and civilian rule. The government rotated between interim presidencies for five years until 1982 when Congress selected Hernán Siles Zuazo to lead the nation (Van Cott 2000a). Siles faced the very difficult and complex task of dealing with years of irresponsible fiscal policies that was compounded by a rise in popular demands for state services, a trend that was unleashed with the end of the military regimes as constituents finally voiced their wants and needs under the new democracy (Van Cott 2000a). For example, organized labor initially gave support to the Siles government but only in return for high raise increases. However, when Siles was unwilling to grant further raise increases, labor turned against him and held strikes that crushed his final attempts at stabilizing the economy in the face of high inflation. While in the Congress, right wing members rejected all attempts to broaden the tax base thereby eliminating the possibility of staving off inflation and lessening foreign indebtedness (Morales and Sachs 1989).

Though Bolivian exports had risen throughout the 1970s, little money had been spent modernizing the mining sector and other industries, and public debt taken on during the 1970s' dictatorships was staggering. By 1984, the servicing of the national debt equaled 36 percent of the value of Bolivia's exports. Though the illegal export of coca provided an informal source of income, it could not substitute for the declining mining economy and lack of international loans. The Siles government saw only one solution, to print more money, and inflation soared further reaching an annual rate of 8,170 percent by the first six months of 1985 (Klein 1992:272). Yet Siles never attacked the large, problem-ridden state enterprises, such as the state mines and petroleum industries. Because Siles was unable to rally support for his stabilization packages, Bolivia faced an economic crisis of unprecedented magnitude marked by the hyperinflation of May 1984 (Morales and Sachs 1989). With popular support waning, Siles agreed to call early elections and stepped down from office in August 1985.

The July 1985 election reinstated the power of MNR, and introduced the right wing party *Acción Democrática Nacionalista* (ADN) headed by ex-dictator Hugo Bánzer. In the end, Victor Paz Estenssoro reclaimed the presidency for his fourth term with support from many factions of society, including much of the peasantry who still associated Paz Estenssoro with the agrarian reform. Abandoning his previous economic policies and economic nationalism, Paz Estenssoro adopted the policies of economic liberalism and within three weeks he constructed and signed into law the New Economic Plan (NEP) (Klein 1992).

The NEP called for liberal economic reforms and a national austerity program to halt inflation (Klein 1992; Morales and Sachs 1989). The

NEP immediately eliminated all price and wage controls, the national currency was devalued, government expenditures were cut back, and payments of foreign debt were temporarily suspended. The result was the quick stabilization of the Bolivian currency. Business investments halted and value-added taxes were imposed so that the country immediately went into severe recession (Klein 1992).

Though the austerity program of the NEP was able to stabilize the economy in a remarkably short period of time, it continues to be debated whether the high costs of austerity in human suffering were worth the payoff in terms of stability (Godoy and De Franco 1992; Jameson 1989; Morales 1991). Organized labor was certainly one of the hardest hit by the NEP, as employment in mining dropped 75 percent by 1986 (Jameson 1989:97). The NEP also did not deal with underdevelopment or poverty, and the only direct beneficiary of austerity policies was the formal private sector (Jameson 1989:98).

Another debate revolves around the effects of inflation and the NEP on agriculture (Godoy and De Franco 1992; Morales 1991). Though Godoy and De Franco (1992) maintain that some agriculture enjoyed "fleeting prosperity" during the inflationary years of 1982–85, they emphasize that this should be analyzed in light of decades of poor agricultural growth. Further, any prosperity in agriculture vanished under the NEP, with smallholder agriculture suffering the most. Government subsidies on transport and credit were eliminated, reduced public employment caused decreased demand for food, while restrictions and tariffs on food imports were abolished. As a result, smallholder agricultural production declined (Godoy and De Franco 1992).

Declining food prices following the NEP are also indicative of declining terms of trade for smallholder agriculturalists. As food prices dropped, the liberalization of food imports simultaneously increased supply, resulting in disastrous terms of trade for smallholders. For example, the price of potatoes fell substantially following the NEP, particularly relative to the price of transportation, which continued to rise causing the terms of trade to decline considerably (Morales 1991:62).

The 1989 presidential elections ended the MNR-ADN alliance as Gonzalo Sánchez de Lozada, the planning minister who had help devise much of Estenssoro's NEP, ran against Bánzer. Following a stalemate between Sánchez de Lozada and Bánzer, Congress finally voted for the third place leftist candidate of the MIR, Jaime Paz Zamora (Klein 1992; Van Cott 2000a).

In 1993, Gonzalo Sánchez de Lozada of the MNR finally won the presidency by capturing a portion of the indigenous vote, and capitalizing on the raising trend in ethnopolitics and themes of multiculturalism that were coming to a head in the late 1980s and early 1990s. The political weight of these movements was confirmed when Sánchez de Lozada choose Víctor Hugo Cárdenas as his vice-presidential running mate and together they won the 1993 elections. Cárdenas was a respected and educated Aymara linguist who represented the MRTKL, one of several indigenous political parties that emerged during the 1970s and 1980s (Klein 1992; Van Cott 2000b).

KATARISMO AND INDIGENOUS POLITICS IN LA PAZ

In the late 1960s and through the 1970s and 1980s, the issues and demands of indigenous Aymara speaking peoples, both the rural peasantry and recent urban immigrants, began to be voiced through numerous social and political organizations. These groups often took the name of the colonial Aymara hero, Túpac Katari, who led the rebellions of 1780–82 in La Paz. The leaders of the *Katarista* movement were university educated urban migrants who had been raised in the rural Aymara heartland of the altiplano. A key to this group's identity was based on its roots in the Aymara peasantry. However, this peasant consciousness was being raised under the challenges, conflicts, and opportunities that these immigrants faced as the first generation of migrants to the city following the 1952 revolution (Albó 1987; Rivera Cusicanqui 1987).

The grassroots difference between the *Kataristas* and other pro-peasant groups that came before them, was that not only were the *Katarista* leaders born into the rural peasantry, but they also continued to maintain direct connections to their home communities and in some cases actually returned to live in the countryside. By doing so, they were instrumental in introducing new political ideologies into the Bolivian countryside, including ethnic and class-consciousness, particularly by becoming involved in the rural peasant unions. Both Raimundo Tambo and Genaro Flores, early leaders in the *Katarista* movement, were born in the countryside and migrated to the city early in life. Once there, both men received a university education, eventually returning to their original home communities (Albó 1987; Rivera Cusicanqui 1987).

However, Flores was to have a special affiliation that aided his career in the peasant unions and in politics. Flores was chosen to be a research assistant in a study led by a team from the University of Wisconsin conducting case studies in rural Bolivian communities (Albó 1987). This allowed Flores to learn more about rural issues and lent him a certain amount of political credentials. Flores was one of several Aymara intellectuals who became involved in politics following an extended affiliation and training under North American or European universities. Included in this group with Flores is Mauricio Mamani, who worked extensively in his home community of Irpa Chico with the North American Anthropologist William Carter and later became a Bolivian Minister of Agriculture. Another *Katarista* leader was Víctor Hugo Cárdenas, who had training as a linguist in Europe and later became Vice-President (Albó 1987)

In the 1970s, the *Katarista* movement led by Flores began as a cultural revitalization movement, later gaining power in the peasant unions, and finally becoming a full-fledged political party in Bolivia by the end of the decade (Albó 1987; Rivera Cusicanqui 1987; Van Cott 2000b). On July 30, 1973 the *Centro Campesino Tupac Katari* led by Flores, and four other groups, signed the Tiwanaku Manifesto that systematically outlined the *Katarista* social, unionist, and political agenda. Among the themes in this document were: demands of respect and dignity for indigenous cultures and values; a denouncement of the exploitation of peasant producers and of the poor terms of trade for agricultural goods; the recognition of a post-revolutionary history of exclusion and co-optation of the peasantry in formal political parties; demands for grassroots representation of the peasantry at the national level in the peasant unions; and an attack on the educational system in Bolivia. The manifesto combined an indigenous ethnic ideology based on discourses of cultural subordination, and integrated it into existing leftist discourses of economic and political oppression (Van Cott 2000b:127).

During the 1970s and 1980s, various *Kataristas* groups employed representations of indigenous rebellion and the moral memory of the rural Aymara as the basis for indigenous ideological discourses. Rivera Cusicanqui (1987:149–50) interprets two recurrent themes in *Katarista* ideology. The first is that *Katarismo* embodied an *Incaic* moral code, similar to that embedded in the Inkarrí myth, that indigenous society was oppressed by a continuing colonial situation that subordinated indigenous peoples. This situation was personified in Túpac Katari as a mythic hero who would return to free the oppressed. The second theme is based on the awareness that indigenous groups in Bolivia currently make up a numerical majority of the population. *Katarismo* represented the return of the Aymara hero who would lead this ethnic majority. This theme is depicted in the recurrent use of the slogan *"Nayawa jiwtxa nayjarusti waranga waranqanakawa kutanipxa"* (I shall die, but I will return tomorrow multiplied ten thousand-fold). Oral tradition attributes these words to the hero Túpac Katari moments before his execution (Rivera Cusicanqui 1987:150, 156).

It was not until after the dictator Hugo Bánzer was ousted, and presidential elections were called, that formal *Katarista* political parties were formed (Van Cott 2000b). The first of these was the *Movimiento Indígena Tupaj Katari* (MITKA), which was the most pro-indigenous in its political platform. This extreme pro-indigenous political ideology was founded on the conviction that the root of all problems for indigenous peoples in Bolivia was due to the Spanish conquest. To ally with the Spanish "invaders" was thus considered useless and perhaps traitorous to the *Katarista* political movement. Therefore, MITKA did not seek to ally with traditional political parties in Bolivia, and proposed to replace the traditional political system with a multicultural state that respected indigenous identity (Albó 1987).

The second party to form during the 1978 presidential elections was the *Movimiento Revolucionario Tupaj Katari* (MRTK). Genaro Flores, Macabeo Chila, and Víctor Hugo Cárdenas led MRTK, which was less exclusionary in its politics then MITKA, though they also shared the opinion that present political parties did not respond enough to peasant needs. The MRTK aspired to a constituency beyond the peasantry, while still guaranteeing that peasant issues were an integral part in its political platform (Albó 1987).

Other factors in the splintering of the *Katarista* movement were the different levels of participation by members of the groups in international indigenous social movements. For example, the MRTK developed its contacts and influence within the union movement and had strong ties with political groups in Bolivia. While the MITKA focused on developing contacts with other indigenous movements throughout Latin America and was involved with international groups that supported pro-indigenous campaigns (Rivera Cusicanqui 1987:137).

The breakdown of the *Katarista* movement into numerous organizations and groups continued throughout the 1980s, so that at one point there were 10 parties with a generalized *Katarista* platform. These various *Katarista* parties enjoyed very strong political and peasant union support (Albó 1994:58). While the Katarista movement was splintering into a number of groups and becoming imbued by financial corruption and political patronage, Albó (1994:61) argues that its main proposals concerning the pluri-national state and the discourses of ethnic awareness were being adopted and embraced in other political sectors

As the neo-liberal reforms of the Estenssoro government in the mid 1980s were systematically destroying the political support bases of the workers unions, literally by dismissing them from their jobs, class-consciousness and the politics of the left fell into crisis. Meanwhile, the dissolution of the former Soviet Union and the ethnic clashes within its former Republics also demonstrated an international "crisis of the left" (Albó

1987). As Albó writes "The crisis of the left, internationally and within Bolivia, led several sectors of the disconcerted left to set their eyes on the peasantry as a possible successor to and alliance with workers and miners, and to take seriously what the *Kataristas* had been saying for ten years" (1987:61). So as the *Kataristas* themselves began to lose internal group cohesion and drive, their indigenous rhetoric and discourse was being taken up by new fronts in the wake of the NEP.

The renewed ethnic consciousness and indigenous rhetoric began to be used in various institutions, organizations, and political platforms. Lowland coca producers, who had already been one of the best-organized and mobilized sectors in the countryside during the 1980s, also began to include ethnic and cultural elements in their discourse. Though for most coca producers the significance of coca on their lives was in its economic value, the main motto was to defend the "sacred coca leaf" by stressing its cultural value (Albó 1994:63). Through this ethnic discourse, coca growers were able to form an alliance with peasant producers and eventually to gain power in the major peasant unions (Healy 1991).

In the 1980s, indigenous groups in the lowlands began to organize opposition to defend their territory and natural resources from loggers who were expanding their economic activities into lowland indigenous territory. This social movement culminated in the August 1990 March for Dignity (Albó 1994; Jones 1990). The March for Dignity had a dynamic impact on the political awareness of indigenous rights for the rest of country. It gave national media coverage to the grievances and issues of both the highland *Kataristas* and the lowland indigenous groups (Van Cott 2000b)

Concurrently, there was a rise in the city of La Paz of what Albó (1994) terms the "*Cholo* Populism" of Carlos Palenque, who began his meteoric rise in the 1980s utilizing ethnic rhetoric and appointing a woman to his parliamentary slate wearing the traditional *pollera* skirt of the rural Aymara. The rise in Palenque's populist campaign culminated when he was elected mayor of La Paz from 1989 to 1991. Through his radio and television shows, Palenque used ethnic discourses and symbols, focusing on people dressed in traditional clothes, speaking Aymara, and appealing to the working classes of the city (San Martín A. 1991).

By the late 1980s and early 1990s, indigenous discourses permeated many disparate sectors of Bolivian politics, sometimes far removed from the political base of the *Kataristas*. The major political parties could no longer ignore the indigenous issues and were prompted to add ethnic laden discourse to their own political platforms. The 1993 electoral campaign and the MNR are a primary example, as Sánchez de Lozada brought on Cárdenas as his running mate and instituted the rhetoric of the pluri-national state into his political platform. The pluri-national state was a form of multiculturalism that stressed the value of cultural diversity within the state, and advocated equality among ethnic groups. Long gone was the rhetoric of the former MNR from the revolution of 1952, which wanted to integrate the *campesinos* into the nation-state by downplaying ethnicity in the rhetoric of revolutionary nationalism. In its place was the new concept of the pluri-national state that had long been the project of Cárdenas and his sector of the *Kataristas*. As Albó (1994:69) observes, the very recognition by MNR to cultural diversity within the state was a long way from the MNR of the 1952 Bolivian Revolution.

Yet the MNR was not the only traditional political party in Bolivia that was trying to appeal to this renewed ethnic and indigenous consciousness. In the late 1980s and early 1990s, MIR and most of the political left also turned towards ethnic discourses as its class base was being dismantled and in crisis under the NEP. Even the right had to pay some heed to ethnic issues (Rivera Cusicanqui 1993), though this rhetoric was not followed up with any solid, long-term moral or material commitment in the ADN coalition government of 1997. As Van Cott writes, "The promotion of ethnic diversity as a major theme of government lost centrality and coherence in the Bánzer government" (2000b:92). By 1997, state discourses on indigenous issues shifted away from multiculturalism and towards an emphasis on reducing poverty. Thus the various movements of indigenous consciousness and ethnopolitics had reached its zenith by the mid 1990s (Van Cott 2000b).

SUSTAINABLE DEVELOPMENT AND INTERNATIONAL DEVELOPMENT POLICIES

In this chapter, I have established the immediate contexts for the raised field project by outlining major economic, social, and political themes in Bolivian society leading into the late 1980s and early 1990s when the raised fields rehabilitation project was being conceptualized and implemented. However, I now turn away from local and national level trends to include issues that were forming in the developed nations of the North who funded and supported the raised field project, both financially and in other ways. Foremost is the advancement of the environmentalist movement and a subsequent turn towards development theories and projects that preached the benefits of "sustainable development." As I will argue in this chapter, the raised field rehabilitation project was represented as a model of the new sustainable development paradigm. In this section, I outline the formation of sustainable development in natural resource management and agricultural development, and link it to growing environmentalist concerns in the U.S. and Europe. I also make the conceptual link between sustainable development and applied anthropology's concern for preserving and promoting the use of indigenous knowledge in development and natural resource management.

The now commonly cited (Adams 1990; Escobar 1995; Gupta 1998) release of the Brundtland Report (Brundtland 1987) marked a paradigm shift in the development agenda towards the practice of what is called sustainable development. Sustainable development is an economic strategy that seeks to promote economic expansion with the least amount of detriment to the environment and natural resources. This section examines this shift away from post-World War II modernization theory and Green Revolution technology in agriculture, towards the new sustainable development of the 1980s and 1990s, specifically by focusing on sustainable agricultural development.

By the 1980s there was a growing recognition within the development community that the tremendous gains in agricultural production under the Green Revolution were in many cases having a deteriorating effect on the environment (Harwood 1990). The Green Revolution introduced new hybrid strains of cultigens that were dependent on chemical fertilizers and pesticides. Unfortunately, traces of these chemicals soon began showing up in the water supplies and underground aquifers in the U.S. (Harwood 1990). In third world nations, the Green Revolution introduced Western styles of agriculture, such as monocropping on extensive tracks of land, which usually included high costs in machinery and fuel that was not within the reach of smallholder farmers in less industrialized nations.

The recognition that many Green Revolution agricultural improvements were having a detrimental effect on the environment, coincided with a renewed focus on global environmental issues (McCormick 1989). Examples of 1970s and 1980s environmental concerns include, the recycling movement in the U.S., global warming, and a general focus on the globalization of the environment, such as the effect of acid rain crossing international boundaries. A wake-up call on the limits of natural resources in the U.S. came during the gasoline shortages of the 1970s, as Americans stood in long lines for gas and realized, perhaps for the first time, that there were limits to this non-renewable fossil fuel. Given these circumstances international development was poised for a fundamental shift in its paradigm and a concomitant change in discourses of development.

The history and origins of sustainable development is closely tied to the environmental movement in the 1960s to 1980s, and is linked to changing concerns over nature in Europe and North America. Key concepts of sustainable development were formed as part of an environmentalism that emerged in the 1960s. Adams (1990) details key environmental concerns that formed part of the groundwork of the sustainable development of the 1980s. The first idea that he outlines was the growing concern for nature and wildlife preservation during the 1960s and 1970s, which influenced the evolving sustainable development paradigm. Adams argues that, "notions of sustainable development were adopted partly as a means of promoting nature preservation and conservation" (1990:16). Thus the desire within the conservation movement to preserve nature and natural resources, combined with the idea that development was having a deleterious effect on the environment, created the push towards new development planning that took into consideration long-term environmental impacts and conservation.

Contributions from the science of ecology were also closely linked to the environmental movement of the 1970s and the sustainable development of the 1980s. The scientific discipline of ecology, according to Adams, "seemed to offer new, value-free, and apolitical ways of not only understanding but also managing the environment" (1990:23). The idea that the environment could be understood and managed through an ecological approach had very strong appeal, since development groups overtly strive to be apolitical or at least to represent themselves as being apolitical. By offering technical and managerial assistance in the fight against world poverty and underdevelopment, ecology seemed to offer a new tool to understand development problems where they had previously failed. For example, tropical ecology had a particularly strong influence on environmentalism and development in the 1980s, as U.S. popular concern grew over the destruction of the complex ecological systems of tropical rainforests in the southern hemisphere. Tropical ecology was closely linked to indigenous social movements in the Brazilian Amazon and drew popular worldwide attention in various crusades for saving the rainforests.

The third conceptual component of environmentalism that underpinned sustainable development is the idea of a global environmental crisis that permeated scholarly works and public concerns starting in the late 1960s and through the 1970s (Adams 1990:27). There was a new global focus on problems, as environmentalists began to recognize that environmental issues were global problems, not merely local or regional problems. By the 1970s the idea that the earth was a complete ecological system had also gained credence in environmental movements so that popular concerns where no longer just about local or even national interests, but spanned the globe. This global awareness of environmental problems caught popular attention as well as scholarly interests (Adams 1990; Harwood 1990).

This global awareness went hand-in-hand with a renewed "neo-Malthusian" perspective that constructed an "apocalyptic vision" of unchecked population growth that was having a devastating effect on the environment. Hence, there was a growing concern for managing population and economic development within sustainable contexts (Adams 1990:28). This neo-Malthusian thinking within certain strains of environmentalism had significant political implications as Northern environmentalists despaired over their Southern neighbors' lack of population control and resulting environmental degradation (Adams 1990:30). Of course, any form of global political intervention by environmentally concerned Northern neighbors would be problematic; therefore environmental issues were addressed through the medium of international economic development. The environmentalist movement's push towards a new sustainable development included the co-management of the environment with modern industrialization. Since development was the key to economic and political control of developing nations (Escobar 1995), it would be through sustainable development that the environmental concerns of the developed nations would be addressed.

The World Conservation Strategy (WCS) (IUCN 1980) was published in 1980, and was the first major development document that attempted to add an element of environmental conservation into the international development agenda. The WCS clearly reflects the environmental program of the 1960s and 1970s with aspects of neo-Malthusian concerns for the global environment. The aim of the WCS was the conservation of natural resources, the preservation of genetic diversity, and "to ensure the sustainable utilization of species and ecosystems . . . which support millions of rural communities as well as major industries" (IUCN 1980: vi). In this report, the authors maintain a clear link between rural poverty, environmental degradation, and the constraints and limits this places on industrial development. For example, the WCS makes a direct link between the rural poor and environmental degradation writing that "hundreds of millions of rural people in developing countries . . . are compelled to destroy the resources necessary to free them from starvation and poverty" (IUCN 1980:vi). On the other hand, the report counters this concern for human welfare and poverty, by stating that "the resource base of major industries is shrinking" which implies that these rural peoples who are being compelled to destroy their own natural resources are also reducing their own capacity for industrialization.

The WCS advocated new mandates for development as an awareness of "global responsibility" and the "interrelatedness of actions," specifically the link between growing rural poverty and the reduced capacity for industrial development (Adams 1990:46–47). Yet Adams (1990) notes that this first attempt at an environmentally friendly development failed on two accounts. First, it did not recognize "the essentially political nature of the development process." Second, it suggested that conservation and the science of ecology could somehow "bypass structures and inequalities in society" (Adams 1990:50–51). Though in its infancy, the WCS brought the environment into the development agenda.

Another attempt to put the environment into the development agenda was the "ecodevelopment" trend of the 1980s. Ecodevelopment first appeared in the 1970s and was promoted by mostly European development institutes (Adams 1990). Within ecodevelopment lies the same ecosystems concept of ecology and a similar emphasis on the need to understand the workings and complexities of ecosystems when implementing development projects (Adams 1990:51–52). Ecodevelopment focused on "basic needs" development that was geared towards satisfying the needs of the "poorest of the poor" in society. It also promoted participatory development with grassroots participation by the subjects of development, called "development from below" (Adams 1990:53). The third focus of ecodevelopment was the promotion of the appropriateness of developing "intermediate technology." This is the "small is beautiful" thinking behind grassroots and participatory development, which played a significant role in agricultural development thinking. This version of development discourse argued that intermediate technology, such as hand tools and other small-scale technology, was more readily available to smallholder farmers, unlike the expensive tractors and other farm equipment. Therefore, a goal of ecodevelopment was to pursue a form of economic development that relied more on human and natural resources, rather than Western capital-intensive technologies (Adams 1990:54).

The release of the Brundtland Report in 1987 was the coalescence of environmentalism, ecodevelopment, and other precursors within the development community towards an environmentally conscious form of development. While the term sustainable development had been bandied about before, such as in the case of the WCS, it was with the release of this report that it finally left the margins of development critiques and took center stage in development discourse (Adams 1990; K. Hamilton 1997). The report both originated and was sponsored by the United Nations General Assembly, and its findings and recommendations had tremendous weight. Its origination in the UN clearly made the link between economic development and international politics more evident than in anytime since the post-World War II economic restructuring period (Adams 1990). The Brundtland Report was able to link development and the environment together as one issue for the international community to debate, inseparable from each other. This was to have a lasting impact on development theory and practice through the 1990s.

At this point it is worthwhile to highlight some of the themes and concerns in the Brundtland Report to illustrate the underlying thinking on sustainable development. Key issues and themes from the Brundtland Report had immediate influence within international politics and the development policy community, and it continued to shape the discourse and objectives of economic develop in the 1990s. The language of "sustainable development" influenced all phases of development discourse after the release of the report, but in no sector of economic development was it to have more influence than in agricultural projects and natural resource management. The ensuing sustainable agriculture school of research and theory has produced a prodigiously large and growing body of literature (Harwood 1990:13–14; Warren 1999).

The Brundtland Report defines sustainable development as "development that meets the needs of the present without compromising the ability of future generations to meet their own needs" (1987:43). The first key concept in this definition of development is that of "needs," particularly the "essential needs" of the world's poor, which is stated to be the major objective of development. These essential needs included food, clothing, shelter, and jobs, which were not being met for much of the world's population. The notion of meeting basic human needs, and of raising the quality of life for the world's poor, resurfaces regularly throughout the report. Like ecodevelopment, sustainable development focused on human social conditions and endemic world poverty.

The concept of poverty is also referenced throughout the report. Poverty is seen as both a cause of environmental degradation and also the object of intense development concern under the new "basic needs" mandate of sustainable development. In the introduction the report states:

> Poverty is a major cause and effect of global environmental problems. It is therefore futile to attempt to deal with environmental problems without a broader perspective that encompasses the factors underlying world poverty and international inequality. (Brundtland 1987:3)

> The Commission believes that widespread poverty is no longer inevitable. Poverty is not only an evil in itself, but sustainable development requires meeting the basic needs of all and extending to all the opportunity to fulfill their aspirations for a better life. A world in which poverty is endemic will always be prone to ecological and other catastrophes. (Brundtland 1987:8)

In these two passages the report does three things: first, it treats poverty as a cause of environmental degradation; second, it constructs poverty as an object of sustainable development under the mandate of meeting basic human needs; and third, it suggests that this poverty can be rectified and is within the capabilities of modern technology and social organization through the practice of sustainable development.

Within the Brundtland Report's definition of sustainable development—that it meet the needs of the present without compromising the ability of future generations to meet their own needs—also lies a concern for the conservation of natural resources such as land, water, and non-renewable resources like fossil fuels. The idea of environmental limits on development is repeated throughout the report, though the focus is less on limits set by the environment itself and more on the limits set by current technology and social organization. This is a subtle but important difference from the extreme neo-Malthusian ideas of earth as a limited resource in earlier environmentalist agendas (Adams 1990:59). The sustainable development of the Brundtland Report places current environmental limits within the contexts of modern technology and socio-economic development.

Conservationist ideas are represented throughout the report, resounding again and again in its emphasis on economic development that does not deplete or destroy ecosystems, plant and animal species, and non-renewable sources of energy. Another environmentalist idea that surfaces in the Brundtland Report is the conceptualizing of the earth as a singular organism, similar to the Gaia theory of the early 1980s (Lovelock 1987). The report states that, "Earth is an organism whose health depends on the health of all its parts" (Brundtland 1987:1), reminiscent of the popular "spaceship earth" concept of the 1970s. This statement reconciles images of interdependence, similar to ecosystems approaches in ecology, with sustainable development practices. The Brundtland Report also harks back to the theme of global crisis and the globalization of the environment, as previous environmental discourses had espoused. However, the report describes not one global crisis, but three "interlocking crises" of the environment, development, and energy (Brundtland 1987:4).

It is clear that the Brundtland Report included environmental concerns in the sustainable development agenda. But how does the sustainable development of the Brundtland Report suggest that we deal with these new problems of ecological and environmental degradation and the concomitant problem of global poverty? The answer of the Brundtland Report is to foster economic growth and further industrial development. However, the report puts this renewed economic growth and development within a political economic context by recognizing that the current terms of trade between developed countries and developing countries are not equitable.[2] Thus the report urges that asymmetrical economic relations be alleviated by increased flows of capital into these regions of the world.

What sets the Brundtland Report apart from the previous WCS and ecodevelopment, is that it does not see sustainable development as simply an assortment of micro-level environmental projects and development planning, but strongly asserts that sustainable development can only be met by international trade and economic growth (Adams 1990:62). The neo-liberal foundations of the report are evident in its criticism of protectionist economic policies, of high debt burdens in developing countries, and of declining monetary support for economic development by the industrialized nations. The Brundtland Report's called for increased capital from institutions such as the World Bank and private lenders, which would fund projects that take the environment into account. The report states that such environmentally friendly projects would usually be smaller projects aimed at enhancing the environment and would include "maximum grass-roots participation" (Brundtland 1987:77). Ultimately, economic growth is viewed as the means to fight poverty, manage the environment in a sustainable fashion, and integrate social and economic development (Adams 1993).

Since the release of the Brundtland Report in 1987, discourses of sustainable development have infused the planning and objectives of economic development projects and development theory (Auty and Brown 1997). For example, the Rio Earth Summit of 1992 issued Agenda 21 that outlined the principles, policies, and programs of environmentally sustainable development, which was endorsed by most countries and international development institutions such as the World Bank (World Bank 1997).

However, both critics and proponents of sustainable development often have a difficult time agreeing on a common definition of exactly what it is and what it is advocating (Auty and Brown 1997; K. Hamilton 1997; Pezzey 1989). Sustainable development has been criticized as not being a very useful analytical concept since its use and meaning is so varied and flexible (Adams 1993). Adams (1993, 1995) refers to sustainable development as something more akin to a slogan rather than a paradigm or basis for theory. He argues that the real value of sustainable development as a concept lies less in its analytical insights, but rather in the way it "links diverse (and sometimes divergent) ideas and blends them, often uncritically, into an apparent synthesis" (Adams 1995:88).

Though the use of the term sustainable development as an analytical concept is varied, I would argue that there are two primary currents in mainstream sustainable development discourses: 1.) focused on ecologically sustainability, and 2.) focused on economic sustainability. Ecological sustainability is directly tied to the influences of the environmental movement as discussed previously (Adams 1990, 1993, 1995; Harwood 1990).

Ecological sustainability focuses foremost on the management of the environment and natural resources. However, the idea of economic sustainability (K. Hamilton 1997) or "livelihood sustainability" (Humphries 1993) has arisen partly in response to the concerns of the environment, but places more emphasis on generating wealth and income, and varies from its emphasis on individual households to the economic stability and growth of the global economy.

Gardner (1997:143–146) maintains that since the 1980s there has been a general shift in the ideals and stated objectives of development towards sustainability, particularly in the U.K., but also in the U.S. For example, anti-poverty and "equity" approaches became more widespread, and income-generation, participation, and grassroots strategies became the buzzwords in late 1980s and early 1990s development discourses. Gender and "women-in-development" was also clearly written into development projects and policies. Development approaches were by no means uniform in this shift towards sustainability and encompassed a vast array of approaches, practices, and activities. For some it still meant intervention to help "less developed" countries towards modernization and economic growth, though for others the focus shifted to a concern for "empowerment," the idea of helping others to change their own lives. It meant "enabling people to meet their individual potential; it was a human rather than an economic process" (Gardner 1997:143). In agriculture, emphasis was placed on "bottom-up" innovations, farmer participation, and indigenous knowledge.

The sustainable development approach in agriculture spawned a whole new sub-field of agricultural studies called "sustainable agriculture" (Gold 1994), which has its own *Journal of Sustainable Agriculture* and several international centers that conduct research and promote sustainable agriculture. Sustainable agriculture clearly reflects the ecological concerns of the environmental movement and attempts to rectify "many of the shortcomings of the Green Revolution model" (Harwood 1990:11).

In one recent publication on sustainable agriculture, Gold (1994:3) defines the goals and objectives of sustainable agriculture. She maintains that sustainable agriculture includes ecological sustainability issues, such as enhancing environmental quality and making the most efficient use of non-renewable resources. It also includes economic sustainability issues, such as sustaining the economic viability of farm operations and enhancing the quality of life for farmers (Gold 1994). Her definition also touches once again on the human needs component of sustainable development. Among the terms that Gold (1994) lists as being commonly associated with sustainable agriculture are: agroecology, biological or ecological farming, integrated pest management that focuses on reduced use of pesticides, low input agriculture to minimize the amount of external off-farm inputs, and regenerative agriculture.

Within sustainable agriculture, Harwood adds that along with efficient resource use, "biological processes within agricultural systems must be much more controlled from within (rather than by external inputs of pesticides), and . . . nutrient cycles within the farm must be much more closed" (1990:15). These ecological components to sustainable agriculture reflect not only environmentalist concerns for conservation of natural resources and ecology, but also a new focus on limiting external inputs. This focus on regenerative agriculture and the limiting of external inputs has its roots in the organic farming philosophy from as early as the 1940s in the U.S. (Harwood 1990).

Another trend in development theory that was intricately tied to the growing focus on sustainable development, particularly sustainable agriculture, was the increasing interest in indigenous knowledge (IK), indigenous knowledge systems, and the use of indigenous knowledge in development. Indigenous knowledge as it is being used in development theory and practice refers to specialized areas of knowledge available to the pre-colonial populations throughout the world. It is often represented in opposition to Western scientific knowledge (DeWalt 1999; Nygren 1999). Indigenous knowledge has also been written about in terms of indigenous technical knowledge (ITK) and as local knowledge. The key to all of these terms is the juxtaposition of Western, scientific knowledge vs. local or native, non-Western knowledge and ways of knowing the world.

A report written by Warren (1991), and published by the World Bank, sums up how indigenous knowledge might be appropriated and used in development. It explores various case studies where indigenous knowledge was ignored by project personnel and resulted in development failures. These were compared to projects that relied on indigenous knowledge during the planning and implementation stages, and which achieved development project success. Proponents of indigenous knowledge have argued for its preservation (Hunn 1999), as well as for use in applied anthropology (Purcell 1998; Sillitoe 1998), in sustainable development (Quiroz 1999) and in natural resource management (DeWalt 1999). Warren (1999) writes that the number of centers focusing on the recovery, documenting, and recording of indigenous knowledge has grow rapidly since 1990 to over 30 centers worldwide, while published literature on the role of indigenous knowledge in agriculture has exploded in the 1990s.

Underlying the conceptual and practical usage of indigenous knowledge in sustainable development is a conservationist philosophy. This conservationism is evident in the underlying idea that Western society must seek to preserve the simple, natural, or wild essence of natural resources and ways of life in non-Western societies. The implication is that nature is somehow being lost in contemporary societies and through modern economic development. In the case of indigenous knowledge, a specific concern is that primitive or pre-modern cultures are being lost and are in need of preservation. The idea is that culture and cultural knowledge must be preserved as modernization, scientific, and/or Western culture supersedes it, and that such indigenous knowledges may hold the answer to immediate and future environmental and development woes.

Yet despite all of the attention given to sustainable development and indigenous knowledge, the use of new development buzzwords, and the focus on poverty, the question remains whether there has been a fundamental change in the structure and practice of global development by developed nations. The answer to this question remains to be debated, while reviews of sustainable development successes and failures are mixed (World Bank 1997). As Adams puts it, "it is far from clear whether sustainable development offers a new paradigm, or simply puts a green wash over business as usual" (1993:207).

However, while it may be debatable how and to what extent sustainable development has changed the practice of development in general, it has without a doubt changed how development is talked about and the images and representations that discourses of development produce. Earlier theories of agricultural development valued mechanization and improved technology. For example, there was the use of hybrid cultigens and chemical inputs under Green Revolution agriculture. While sustainable development tends to value small-scale projects, appropriate technology, and organic/regenerative agriculture.

Yet sustainable development discourses too often rely on polemic and oppositional categories. Instead of grand large-scale development projects using the latest technology and involving the most knowledgeable scientists and science from Western societies, sustainable development seems to fall into the role of representing the opposite of these traits. Thus, sustainable development has come to represent indigenous efforts and small-scale participatory projects. It is equated with traditional practices and indigenous self-sufficiency (Roseberry 1993:328). Yet these representations and definitions of sustainable development have often been applied uncritically and in too broad and encompassing a fashion so that anything that is not modern is considered sustainable and worth preservation. This broad and uncritical use of the term sustainable development, and all the associated practices that go with it, will be examined in the following chapter through the use of a case study of the raised field rehabilitation project in the Bolivian Lake Titicaca Basin.

CUSTOM DESIGNING DEVELOPMENT: AN ALTIPLANO DEVELOPMENT PROJECT IN BOLIVIA

To conclude this chapter, I propose that several convergent, interconnected, and dynamic processes were taking place in Bolivia, and in the international development community, that laid the foundation for the raised field rehabilitation project. These convergent themes formed the characteristics for a new development agenda on the Bolivian altiplano. The characteristics of an appropriate and potentially successful development project included small-scale projects, using indigenous technology, and relying on manual labor and local natural resources. I argue that it is no coincidence that these trends came to be considered appropriate for an altiplano development project, but were in fact directly linked to the social, political, and economic trends that I have outlined in this chapter.

The mid 1980s were marked by severe economic recession throughout the country with continuing declining terms of trade for agricultural goods. The harsh government structural adjustments imposed under the NEP cut what little benefits and jobs that had been available to the urban working classes, while the living conditions for the urban and rural poor continued to deteriorate. In the wake of the NEP, there were few funds and little interest in developing costly and highly mechanized rural economic development projects for the altiplano, especially considering that most rural development at that time was directed towards the tropical lowlands.

During the late 1980s and into the 1990s, government interest and investment in agriculture and rural development tended to be directed towards the colonization and development of the eastern lowlands, and particularly in activities aimed at decreasing the cultivation and production of coca in the Chapare region. On the altiplano, there was little government interest in developing subsistence agriculture that rarely produced a surplus for the markets of the cities. Following neo-liberal reforms focused on self-reliance, decentralization, and privatization, NGOs began to increase in number in La Paz and formed an intermediary link between the government, international sponsors, and the rural communities.

By the 1980s, the *Katarista* movement was also beginning to have a significant influence in Bolivian politics and national consciousness. In national politics, ethnic and indigenous discourses infiltrated all political parties and political platforms to some extent by the end of the 1980s and early 1990s. This focus on ethnicity and indigenous culture was included in political party platforms, though the sincerity of this ethnic inclusion varied a great deal between parties. Regardless, the nation as a whole seemed to take a growing national pride in its indigenous heritage and co-opted the ethnic flavor of the *Katarista* movement.

Meanwhile, development discourse, theory, and practice were undergoing some heady changes, primarily as a direct result of the environmental movement that was shaping the politics of Northern hemisphere industrialized nations. Sustainable agricultural development promoted small participatory projects that included local and indigenous farmers in planning and implementation. Such projects often sought to include local peoples and their cultural practices through projects that espoused indigenous self-development and the use of indigenous knowledge. Sustainable development often promoted the use of intermediary technology rather than Western mechanization, and the use of on-farm inputs such as natural animal manure and manual labor. Coincidentally, these small projects seemed to be a perfect fit with the small indigenous communities of the Bolivian altiplano in the minds' eye of development personnel. The fact that neither the state nor international sponsors cared to invest large sums of capital into developing subsistence farms (rather than the market oriented production at lower altitudes) went hand-in-hand with the strategy of low external inputs of sustainable agriculture.

Meanwhile, the indigenous knowledge focus within sustainable development also struck a cord with the renewed indigenous political discourses in Bolivia. The focus on indigenous knowledge in models of sustainable development fit well with the *Katarista* agenda of valuing and elevating Aymara culture. Therefore, when the raised field rehabilitation project proposed to redevelop an ancient indigenous knowledge of the Aymara, it held a certain appeal for urban Bolivians who were reconsidering multiculturalism and the aspirations of a pluri-national state.

These multiple and inter-woven trends in the late 1980s and early 1990s paved the way for the raised field rehabilitation project. As I argue, these trends established the contexts for rural development, and rural development projects, on the Bolivian altiplano in the late 1980s. The uncanny fit between the characteristics of what was considered an appropriate and potentially successful development project, and the raised field rehabilitation project, made it appear as though this project had been custom designed for the Bolivian altiplano. As I will illustrate in the following chapter, the representation of raised fields was systematic and thorough in its depiction of the rehabilitation project as ecologically sustainable, appropriate technology, and indigenous knowledge.

NOTES

[1] *Pongueaje* service is a labor service obligation owed to large estate owners in return for access to land.

[2] The Brundtland report states, "economic partners must be satisfied that the basis of exchange is equitable; relationships that are unequal and based on dominance of one kind or another are not a sound and durable basis for interdependence" (Brundtland 1987:67).

CHAPTER FOUR

"Inventing Tradition" and Development: The Representation of Raised Field Agriculture

I SSUES OF CULTURAL REPRESENTATION HAVE BECOME A THEME IN anthropology over the past three decades (Kellogg 1991; Marcus and Fischer 1986; Ortner 1984). Following Said's (1978) groundbreaking work, anthropologists have examined the ways that we represent other cultures through our work and in our ethnographies (Clifford 1983; Rabinow 1986). Anthropologists working in the Pacific culture area have focused attention on cultural representations of tradition and custom, by using history as an analytical tool to view how such traditions and customs originated (Carrier 1992a). Likewise, the concepts of tradition and custom have caught the interest of nation-states attempting to regulate culture through cultural policy (Lindstrom and White 1994).

Continuing with the theme of tradition, Hobsbawm and Ranger (1983) explore the ways that traditions have been invented and instituted in an attempt to establish continuity with the past. In exploring the invention of traditions, researchers have focused on traditions that were invented by dominant ethnic groups and nation-states. However, less attention has been paid to the ways that archaeology interprets and reconstructs the past, and how the interpretations of archaeology are being utilized by contemporary societies and ethnic groups who wish to appropriate and lay claim to the past. Recent research by cultural anthropologists has begun the task of exploring the multiple ways that archaeology represents the past and is integrated into contemporary discourses of identity, knowledge, and values (Benavides 2000; Smith 2000).

In the previous chapter, I outlined several trends in Bolivia and globally that shaped the contexts for development in highland Bolivia during the late 1980s and early 1990s. First, Bolivian altiplano agriculture was

79

subsistence oriented for several interrelated reasons including declining terms of trade in agriculture and government structural adjustments that imposed fiscal austerity measures. Second, the Aymara *Katarista* social movement that emphasized indigenous identity influenced political discourses and was eventually appropriated by the established political parties in Bolivia. Third, there was a fundamental change in the discourse and the practice of international economic development policy towards environmentally and economically sustainable projects. This sustainable development model emphasized projects that were smaller in size, used participatory frameworks for implementation, and were aimed at reducing poverty and empowering the poor and underprivileged in societies.

The culmination of these trends set the historical contexts for the raised field rehabilitation project in Bolivia. Archaeological investigations into the pre-Hispanic raised fields of the Lake Titicaca Basin began in the 1980s, led by the North American archaeologists Clark Erickson in Peru and Alan Kolata in Bolivia. These initial archaeological investigations with raised fields were aimed at excavating and reconstructing the fields for the purpose of understanding the extent of their use and economic importance in pre-Hispanic civilizations. However, in both Peru and Bolivia, enthusiasm over the prolific production on the raised fields soon led to interest by development NGOs in rehabilitating this form of agriculture for contemporary farmers.

This chapter examines the representation of the raised fields, and the raised fields rehabilitation project, from the archaeological experiments that were carried out in Bolivia and Peru during the 1980s, through the full-scale development project in the Bolivian Lake Titicaca Basin in the early 1990s. I argue that the Bolivian raised field rehabilitation project was an attempt to implement an invented tradition in altiplano agriculture. Thus, this work has the unique perspective of documenting the process through which an invented tradition is created. I will illustrate how raised fields were represented, and argue that the thorough and systematic representation of raised fields as indigenous knowledge was an invention tradition.

I explore the multiply ways that raised fields were represented in reports and publications produced by professional archaeologists and by the Bolivian NGO that carried out the development project. I discuss representations of raised fields from interviews with the NGO personnel and with participants in the rehabilitation project. In general, I found that raised fields were represented as an indigenous knowledge, and as a form of sustainable agriculture that was ecologically and economically appropriate to the Lake Titicaca Basin. In discussing the raised fields, archaeologists and NGO personnel used a lexicon based on the sustainable development paradigm to define the raised fields as sustainable agriculture, appro-

priate technology, and indigenous knowledge. Ultimately, these essentialized representations of the raised field rehabilitation project as an invented
tradition failed to engage the imagination and long-term support of the
indigenous farmer-participants.

ESSENTIALISM, REPRESENTATION, AND INVENTED TRADITIONS

In the 1980s, the discipline of anthropology and cultural anthropologists in
particular were enmeshed in an internal debate that questioned the very
foundations of ethnographic inquiry. Following Said (1978), anthropologists began to consider how anthropologists represent other cultures in our
ethnographies and to explore the political and discursive influences that
inform our interpretations of other cultures (Clifford 1983; Clifford and
Marcus 1986; Marcus and Fischer 1986; Rabinow 1986; Trouillot 1991).
This questioning of cultural representations of other cultures prompted
anthropologists to redefine the practice of anthropology (Marcus and
Fischer 1986;0 Ortner 1991) and redefine some core conceptual tools such
as the concept of culture (Abu-Lughod 1991).

It can be argued that the political uses of archaeological representations of the past is equally as important to consider as cultural anthropologists' representations of other cultures in their ethnographies. For one reason this is because the product of archaeological research, and the representations of past cultures that this research produces, often has direct and
significant consequences for contemporary ethnic groups. Ethnic groups
that appropriate archaeological representations often do so to simultaneously differentiate themselves from other groups and to authenticate their
own group status by laying claim to the past. Yet archaeologists have paid
less attention to this anthropological debate, though some have begun to
consider the politics of representation in their work (Gathercole and
Lowenthal 1990; Patterson 1999; Schmidt and Patterson 1995).

One problem with the representation of other cultures, both past
and present cultures, is that often these representations have been tempo-
rally stagnant and essentializing of the subject that they are representing. In
Said's (1978) portrayal of Orientalism he argues that the representation of
the Orient by Western scholars and travelers was an essentialized image
consisting of a group of traits or characteristics that represented the Orient
as categorically opposite of the essentialized image of the West. Therefore,
the problem with essentializing cultural representations of other cultures
lies in questioning who has control over how a culture or ethnic group is
represented (Dirks, Eley, and Ortner 1994).

The problem concerning power and cultural representation is that
those who have the power to construct and/or manipulate cultural repre-

sentations, also inform and reproduce a sense of social reality and group identity. For example, Said (1978) describes how Orientalist discourse systematically and symbolically represented the Orient as oppositional and subordinate to Western culture and society. His argument is that through cultural representations, dominant groups are able to maintain inequality and the imbalance of power. If this is the case, than how cultural anthropologists and archaeologists represent other cultures might be questioned as a form of colonialism that maintains the subordinate status of other cultures.

Many have added to Said's analysis of cultural representations. For example, Carrier (1992c) argues that Orientalism is not merely a reified and imposed representation from the West onto another culture, but it is a dialectical process of constructing identity between the Orient and the West. The construction of identity through cultural representations takes place within two-way exchange between two groups. Therefore, for Carrier (1992c), the key to understanding the essentializing elements and power evoked in cultural representations is in understanding the history of inequality for each set of actors involved and the group being reified, as well as the history that defines the relationships between them.

Nicholas Thomas (1992) adds a further caveat, that when cultural practices are essentialized they have the potential to be used in different and more political ways. According to Thomas, through a process that he calls "substantivization," cultural practices become institutionalized into a single, essentialized form. Therefore, not only does the group become more clearly differentiated from other groups, but they are also made more cohesive internally as well. Nordholt (1994) gives an example of this process in his article illustrating how Dutch administrators, anthropologists, and other outsiders to Bali, came to view "traditional" Balinese culture as a single representation based on an image of the archetypical Balinese person. This representation was constructed under colonial rule and became accepted and reproduced by the colonial government, but also continues to be reproduced among contemporary local administrators. Where previously there may have been a broader range of practices, through substantivization there becomes a single standard and institutionalized practice, and the creation of a single archetypical Balinese culture

I argue that the representation of the indigenous peoples and technology associated with the raised field rehabilitation project must be understood through an analysis of contemporary social relationships between the subjects (the indigenous farmer participants) and the authors of this representation (the archaeologists and development workers). The representations of raised fields and indigenous people maintain an essentialized image of the local farmers participating in the project as indigenous and ancestral to the region, while also defining their subordinate status in Bolivian society. These contemporary representations of a timeless indigenous technology and indigenous peoples, maintains social distance between the Aymara speaking farmers and urban mestizo Bolivians, and also between the farmers and North American archaeologists.

In Escobar's (1995) seminal work on discourses of development, he examines issues of power and cultural representation in the economic development of the Third World since World War II. Escobar (1995) argues that the concept of development and the idea that any world area needed to "be developed" was primarily an invention of post-World War II elites in the West. His argument is that the "third world" was produced and represented through a discourse of development. The development discourse that Escobar explores "created a space where only certain things could be said or even imagined" (1995:39). Thus the root of development discourse for Escobar was the ideology that modernization was the only course of action for promoting economic progress. Underlying this ideology of modernization and development, industrialization and urbanization were viewed as an inevitable and necessary mechanism of economic progress. For example, Escobar (1995) illustrates how representations of the "third world" and "poverty" became dominant in development discourse and shaped the social reality that was the foundation of development policy. Such representations of the third world continues to be depicted in the discourses of sustainable development in the 1980s and 1990s, where emphasis is placed on developing the indigenous, traditional cultures, and traditional forms of knowledge.

As I have illustrated in the previous chapter, models of sustainable development were often linked to indigenous knowledge and indigenous peoples. As I will demonstrate in this chapter, the discourse of the raised field rehabilitation project also produced representations of raised fields as indigenous knowledge. This representation of raised fields as indigenous knowledge corresponded with contemporary discourses of sustainable development.

I argue that the raised field rehabilitation project was an invented tradition for the Aymara of the Bolivian Lake Titicaca Basin. Hobsbawm and Ranger (1983) addressed the issue of invented traditions and explore how certain "'traditions' which appear or claim to be old are often quite recent in origin and sometimes invented" (Hobsbawm 1983:1). Hobsbawm defines invented tradition:

> 'Invented tradition' is taken to mean a set of practices, normally governed by overtly or tacitly accepted rules and of a ritual or symbolic nature, which seek to inculcate certain values and norms of behaviour by repetition, which automatically implies continuity with the past. In fact, where possible, they normally attempt to establish continuity with a suitable historic past (Hobsbawm 1983:1).

Thus, through the symbolic connection of raised field agriculture with indigenous peoples and the pre-Hispanic past, archaeologists and development workers maintained a cultural representation of rural Aymara speaking peoples as indigenous, and therefore as subordinate.

In the case of the raised field rehabilitation project, I argue that the rehabilitation of an agricultural tradition of raised fields by urban Bolivians, both Aymara and non-Aymara, embodies a discontinuity between the creators of these representations and the cultural traditions of the peasantry that raised fields are representing. There is a discontinuity between the essentialized representations of raised field agriculture and the Aymara speaking peasantry that would practice it, and the urban Bolivian development group who created and implemented it. This discontinuity reinforces and maintains the boundaries between the rural indigenous peasantry and the urban mestizo middle and upper classes.

ARCHAEOLOGY AND EXPERIMENTS WITH RAISED FIELD AGRICULTURE

Raised fields are a labor-intensive method of agriculture once practiced throughout the Western hemisphere. Pre-Hispanic remnants of the fields are found in North America, though they are more widely distributed throughout Central and South America. The pre-Hispanic raised fields occur in various climatic zones ranging from the South American humid tropics to the temperate zones of the North American upper midwest (Kolata et al. 1986). Also known as drained fields (Darch 1983), raised fields are a method of wetlands management that puts seasonally inundated land into cultivation (Denevan, Mathewson, and Knapp 1987; Farrington 1985; Smith, Denevan, and Hamilton 1968). Remnants of prehistoric raised fields have also been documented in Surinam and Oceania, especially in highland and lowland Papua New Guinea. In both New Guinea and Mexico, the ancient raised field systems are still in use (Turner and Denevan 1985). In the Basin of Mexico, pre-Hispanic platform agriculture (known as *chinampas*) has practically disappeared, though they were once the famous and elaborate "floating gardens" of the ancient pre-Hispanic Aztec capital (Chapin 1988).

Extensive excavations and exploratory research in the Lake Titicaca Basin of Bolivia and Peru during the 1980s and 1990s intended to explore the history of raised field use (Erickson 1987, 1988a; Graffam 1990, 1992; Kolata 1993), how they functioned (Biesboer et al. 1999; Carney et al. 1993, 1996; Kolata and Ortloff 1989, 1996; Ortloff 1996; Ortloff and Kolata 1989; Sánchez de Lozada 1996), their social organization of production (Erickson 1993; Erickson and Brinkmeier 1991; Garaycochea 1987; Kolata 1993; Seddon 1994), and the production factors for contemporary cultivation on the fields (Kolata et al. 1996; Sánchez de Lozada 1986). As experiments in raised field agriculture began to produce impressive initial yields of potatoes and other native crops, archaeologists and other social scientists began to investigate and promote raised fields as a possible alternative agricultural development strategy (Erickson 1985, 1988b, 1992a, 1992b; Erickson and Brinkmeier 1991; Erickson and Candler 1989; Garaycochea 1987; Kolata 1996a, 1996c; Kolata et al. 1996).

Raised fields elevate the planting surface of the fields, so that plants are raised above the water level in seasonally inundated areas of the lake basin. These seasonally inundated lands often have not been cultivated in recent history, many of which have lain fallow and been used as pastures for animals since colonial periods and earlier. By reconstructing the raised fields, researchers have demonstrated that the raised beds were surrounded on all sides by canals that capture run-off water and retain solar heat to protect the fields from frost. In the Bolivian Lake Titicaca Basin approximately 30–60% of any segment of raised fields is comprised of planting surfaces, with the rest of the land area in canals that catch the run-off water (Kolata and Ortloff 1996). The canals also help to retain topsoil and resist field erosion (Carney et al. 1996), which is a problem on hillside fields. The standing water in the canals creates a microclimate effect by retaining solar energy in the form of heat that is released during the night (Kolata and Ortloff 1989, 1996). In theory, this heat storage helps to protect the fields from frost in a region where regular killing frosts during the growing season poses the greatest threat to agriculture. However, since these fields have often lain fallow for generations and even centuries, the increased nutrient base of the virgin soils would probably have a similar effect of increasing the plants' level of resistance to frost.

Another aspect of raised field cultivation is the creation of "green manure" in the standing water of the canals that surround the raised beds (Erickson 1988b). The anaerobic conditions in the standing water around the platforms creates this green manure, which can be used to self-fertilize the fields by transferring the canal muck to the field bed (Biesboer et al. 1999). Some researchers hypothesize that the annual application of the green manure produced by the canals, in addition to the reapplication of the topsoil that has eroded into the canals, allows the raised fields to be planted annually without having to lay in fallow for long periods of time like typical dry fields and hillside fields (Erickson 1988b, 1992b).

Contemporary conventional dry fields and hillside fields must be rotated between different crops, followed by a fallow period. In a typical crop rotation, potatoes are planted first for one year, and followed by a rotation of other crops such as *quinoa* and barley, and finally ending in a fallow period. Under conventional cultivation on the altiplano, potatoes are almost never planted more than one year without requiring a long fallow period to allow the soil to rest. At the high altitude of the Lake Titicaca Basin, potatoes are the staple crop, with additional crops such as other native tubers, the native chenopod *quinoa,* and the introduced crops of barley and wheat rounding out agricultural production. Due to the severity of the cold, maize cannot be grown, which is a staple crop in other areas of highland Latin America. Therefore, the claim that raised fields can be planted annually without the need for fallowing the fields is a significant aspect of their potential agricultural production. However, this claim has not been proven and research with communities that participated in the rehabilitation project did not attain much success in planting any fields in potatoes for more than two seasons.[1]

THE RAISED FIELD REHABILITATION PROJECT IN BOLIVIA

The idea for the Bolivian raised field rehabilitation project was a direct result of the experimental raised fields that were built by archaeologists in Bolivia and Peru in the mid 1980s. Archaeologists Alan Kolata from the University of Chicago and Oswaldo Rivera from the Bolivian National Institute of Archaeology built the first raised fields in Bolivia in the agricultural season of 1986–87 (Kolata 1996c). On the Peruvian side of the lake, archaeologist Clark Erickson and his team planted their first experimental fields as early as 1981 (Erickson 1988a). Both teams of researchers met phenomenal initial success in their first year building and cultivating the experimental raised fields. By all accounts, the Bolivian raised fields built during the 1986 season in the community of Lakaya in the Catari Valley produced an impressive potato crop that first year. As Kolata writes:

> In the end, the potato yields on the raised fields averaged over fifteen metric tons per hectare, ten times the yield on the traditional, dry-farmed fields, and all without the use of fertilizers. Some individual plots of land gave as much as twenty-seven metric tons of potatoes. All of the other crops we planted in Lakaya that first year performed just as well as the potatoes. Winter wheat, barley, oats, beans, onions, carrots, and even lettuce weathered the frost and yielded bumper crops. The amazed expressions on the faces of our raised-field farmers were priceless (Kolata 1996c:256).

These phenomenal initial yields inspired the Bolivian co-director of the archaeology project, Oswaldo Rivera, to form an NGO that would rehabilitate raised field farming on a larger scale throughout the Bolivian Lake Titicaca Basin. The NGO *Fundación Wiñaymarka* was a direct result of the archaeological investigations and experiments of Kolata and Rivera. At that time, Rivera was the director of the National Institute of

Archaeology in Bolivia (INAR). The development NGO *Fundación Wiñaymarka* garnered international support and was able to raise funds from organizations such as *PLAN Internacional,* the Inter-American Foundation (IAF), and the Technical Mission of Holland (COTESU).

The objectives of the raised field rehabilitation project funded by the Inter-American Foundation were:

1. to catalyze the incipient movement toward permanent, enhanced yield, raised field cultivation by local communities surrounding Lake Titicaca.
2. to initiate an experimental program of rehabilitating a second form of traditional agriculture in the altiplano: terrace agriculture.
3. to intensify scientific research into the ecological bases of sustainability of raised field agriculture, generating a technical assessment of the feasibility of expanding this form of cultivation throughout the Bolivian altiplano and other regions (Inter-American Foundation 1991).

Specifically, the NGO *Fundación Wiñaymarka* was formed to "investigate and promote the application of systems of indigenous knowledge and traditional values in projects of local cultural and economic development" (Inter-American Foundation 1991). While the definitive goal of the project was, "to stimulate adoption of enhanced yield raised field agriculture among chronically poor rural villages" (Inter-American Foundation 1991). The goals and objectives of the Bolivian raised field rehabilitation project meshed well with the mandate of sustainable development and the promotion of indigenous knowledge. As I will illustrate, these objectives also meshed well with the stated goals of the U.S. based Inter-American Foundation who was the primary funder of the rehabilitation project.

Following the successful experiments cultivating raised fields in 1987 in several communities in the Catari Valley, the NGO *Fundación Wiñaymarka* was able to secure over $1,000,000 in funds for the reconstruction of raised fields in the early 1990s (Inter-American Foundation 1991). The primary funding agency of the rehabilitation project in Bolivia was the Inter-American Foundation (IAF). IAF is a foundation supported by the U.S. Congress and the Social Progress Trust Fund. Among its development strategies IAF lists two goals: the promotion of resource mobilization and local development. Resource mobilization is a strategy based on the expansion of partnerships with Latin American corporations, banks, philanthropic organizations, and public institutions, "with the goal of mobilizing both local and international resources for development" (Inter-American Foundation n.d.). Local development takes this resource mobilization strategy to the local level as a form of "grassroots development" (Inter-American Foundation n.d.).

IAF promotes itself as a conduit for mobilizing grassroots development projects, though they do still fund a number of smaller sustainable development projects through their Local Development Program. The mission of IAF's Local Development Program is:

> Improve the quality of life of the poor in Latin America and the Caribbean by promoting innovative alliances and partnerships involving civil society and the public and private sectors, which result in the implementation of social and economic development projects reflecting local needs and priorities (Inter-American Foundation n.d.).

Other features of the Local Development Program include an emphasis on "participatory" and sustainable development. Sustainable development for IAF is defined as "the durability or permanence of social and economic activities that improve and sustain quality of life, foster economic growth, preserve the environments, and properly manage renewable resources" (Inter-American Foundation n.d.). As I have illustrated in the previous chapter, it is easy to see how the goals and objectives of both the NGO *Fundación Wiñaymarka* and the Inter-American Foundation paralleled the goals and objectives of the sustainable development model of economic development.

With most of its funds from IAF, the budget of the NGO *Fundación Wiñaymarka* included salaries for a director (Rivera), a coordinator/administrator, an accountant, an agronomist, and a logistic coordinator. Other staff members also worked in the local communities, teaching about raised field cultivation and supervising their construction throughout the Bolivian Lake Titicaca Basin, including the Catari and Tiahuanaco Valleys. In fact, salaries accounted for over 46% of the budgeted $1,017,870 in the years that the project was supported by IAF.

Throughout the late 1980s and early 1990s, more and more communities in the Bolivian Lake Titicaca Basin became involved with the raised field rehabilitation project. In 1988, the first raised fields were built in the Tiahuanaco Valley in the community of Achuta Grande. Achuta Grande borders both the community of Wankollo to the west and the town of Tiahuanaco. At about the same time, the NGO *Fundación Wiñaymarka* provided extension services and education in the region by teaching a course on raised fields in the town of Tambillo for community leaders from the provinces of Ingavi and Los Andes. This course taught about raised field construction and cultivation methods.

In 1988, with the help of the Bolivian Army, raised fields were constructed in a community directly north of the ancient ruins at the epicenter of the pre-Hispanic site of Tiwanaku (Kozloff 1994). Standing at the height of the Akapana pyramid, one has an excellent view of the reconstructed raised fields, which complements the lecture given by the tour guides on the ancient agricultural system.

In 1989, *Wiñaymarka* constructed more raised fields in Achuta Grande with the help of the local community, and expanded into other communities in the Tiahuanaco Valley, including the first small test plot in the community of Wankollo. By the 1990 and 1991 agricultural seasons, *Wiñaymarka* had expanded construction of raised fields to a total of 8 communities in the Tiahuanaco Valley, including large-scale community fields and smaller family plots in Wankollo. By 1990, *Wiñaymarka* was in full-scale development and rehabilitation of the raised fields. After 1991, *Wiñaymarka* began to focus on developing the raised fields in other areas outside of the original pre-Hispanic Tiwanaku agricultural heartland, expanding production into areas around the contemporary towns of Batallas, Pukarani, and Copacabana.

The normal method for garnering participation from communities is illustrated in the following outline. First, *Wiñaymarka* approached community leaders and described the raised fields and the potential advantages of this method of cultivation. The community leaders then took this information and disseminated it to his or her respective community. The community as a whole would collectively decide if they wanted to participate in the raised field project (Kolata et al. 1996:207). Communities that wished to participate in the project would formally solicit the NGO to participate in the project. *Wiñaymarka* then provided a raised field instruction course that taught about the cultural origins of the fields, its constructions, and the benefits of this method of agriculture (Kozloff 1994). *Wiñaymarka* also provided pamphlets on raised field construction, cultivation, and production for the participants in the course. Kolata et al. (1996) write that most community members did not seem to reference these manuals very often during construction and cultivation (perhaps because most adults living in the rural countryside are functionally illiterate). A few members from each community would also go on a short field trip to nearby communities that had built raised fields so that they could see the "successful" fields that were already in production (Kolata et al.1996:208). The staff of *Wiñaymarka* would then assess the community's land for raised field construction, determining if there were remnants of ancient fields that could be rehabilitated or if new fields must be designed (Kolata et al. 1996; Kozloff 1994).

Following the site visit, the community signed a formal agreement to work with *Wiñaymarka* (Kolata et al. 1996). The standard agreement with the NGO was for the communities to receive potato seed for the first year of cultivation, basic hand tools including picks, shovels, mattocks, and wheelbarrows, and technical assistance for reconstructing, planting, and maintaining the fields (Kolata et al. 1996). In the early years of the project, communities also received various types of food (Kozloff 1994). I found that this was also the case for participants of the project from the community of Wankollo, who received cooking oil, flour, and a variety of food items for participating in the project. The "food-for-work" incentive of the project was typical on the altiplano and somewhat expected of any project that wanted to commence work in the area. Such an incentive decreased the risk for participants in the project should production on the fields fail. In return, the communities agreed to give a portion of their potato harvest back to the project in order to replenish the NGO's seed stock for cultivation the following year.

By 1991, over 30 communities were building and cultivating raised fields, although some of the earlier communities had already discontinued cultivation on the fields. At that time the total area of rehabilitated raised fields in Bolivia built by *Wiñaymarka* was nearly 60 hectares and included approximately 1600 participants (Kozloff 1994). According to *Wiñaymarka,* by 1992 the number of communities participating in the raised field rehabilitation project had risen to 52, and the total number of individual participants had reached approximately 2,500 people (Kolata et al. 1996:207). Clearly, the early 1990s were the heyday of raised field cultivation under the guidance of the NGO *Fundación Wiñaymarka* and the raised field rehabilitation project.

REPRESENTATIONS OF TIWANAKU AND THE LAKE TITICACA BASIN IN ARCHAEOLOGY

It is sometimes difficult to draw the line between the archaeological literature and the development literature in regards to the Lake Titicaca Basin raised fields, because often it was the archaeologists themselves who proposed rehabilitating raised fields in contemporary communities (Erickson 1988b; Erickson and Candler 1989; Kolata 1996c; Kolata et al. 1996). However, in this section I outline several themes in the archaeological representations of the Lake Titicaca Basin, the pre-Hispanic Tiwanaku State, and past and present inhabitants of the Lake Titicaca region. I primarily focus my analysis on Bolivian archaeology at Tiwanaku and in the southern Lake Titicaca Basin.

In the introduction to an edited volume, Alan Kolata gives an excellent review of the history of Tiwanaku Studies (1996c:4–7). In his review, Kolata describes in detail how early European travelers and writers, and later archaeologists, represented the pre-Hispanic site and civilization of Tiwanaku. Kolata maintains that Tiwanaku Studies have been dominated by representations of the Lake Titicaca Basin as a marginal physical environment not conducive to intensive agriculture. Based on this representation of the Lake Titicaca Basin as unproductive and inhospitable, early researchers interpreted the site of Tiwanaku as not a "true city" or "seat of dominion" for a pre-Hispanic polity. Kolata argues that early researchers tended to view the site as a sparsely populated ceremonial center, a focus of religious pilgrimage perhaps, but not a vibrant and densely populated urban center.

Kolata (1996a) maintains that representations of the pre-Hispanic civilization of Tiwanaku and the physical environment of the Lake Titicaca Basin, which depict the area as barren and the archaeological site as a mere ceremonial center, continues to play a role in the work of some contemporary archaeologists and ethnohistorians. Implicit in such interpretations is that the Lake Titicaca Basin could not have supported a large human population based on local intensive agriculture, with the corollary assumption that Tiwanaku was an isolated collection of architecture devoid of a large resident population. This is the legacy that Kolata and the group of researchers affiliated with his archaeological project faced when they began work in the late 1970s and early 1980s.

The collection of articles in *Tiwanaku and its Hinterland* (Kolata 1996b) is the result of nearly 20 years of research by Kolata and others affiliated with his project in the Bolivian Lake Titicaca Basin. As Kolata writes:

> The data and interpretations presented in this volume demonstrate empirically that the perception of the high plateau as a marginal environment for human production is distorted and ultimately incorrect. Although the plateau lies at an altitude greater than 3,800 meters above sea level, near the upper climatic boundary for viable agriculture, its presumed "marginality" is an illusory concept derived more from anecdotal assumptions about what an agriculturally productive environment should look like than from systematic, empirical research. A principal conclusion of this volume is that, given certain climatic parameters and the appropriate technology and organizational bases, the Andean high plateau offers an enormously rich environment for sustained, intensive agricultural production and therefore the potential for supporting dense concentrations of humans (Kolata 1996a:7).

The collection of articles in *Tiwanaku and its Hinterland* is clearly trying to debunk the earlier preconceptions, assumptions, and representations of Tiwanaku and the Lake Titicaca Basin. The debunking of earlier misguided and ethnocentric representations of the region, and the peoples who once lived there, is certainly a large step in the right direction for Andean archaeology in the lake basin. For example, in chapter two Binford and Kolata (1996) describe the altiplano region of the southern lake basin as not nearly as harsh an environment as it "appears from casual observation" and as early researchers and some current development workers in the lake basin portray it. They write, "the altiplano, particularly in the Lake Titicaca-oriented sustaining area of Tiwanaku, supplies many resources that are readily extracted and processed for human use" (Binford and Kolata 1996:54). They add to these conclusions that "despite superficial appearances to the contrary, the altiplano is potentially a resource-rich, productive environment for complex human activities, *given the development of appropriate exploitative technologies*" (ibid.). Binford and Kolata maintain that the altiplano is still a risky agricultural environment and that even slight changes in the climate of the region can spell the difference between prosperity and agricultural decline.

Binford and Kolata also accuse other Western-trained scientists, particularly those associated with development projects, as routinely perceiving the altiplano as a "marginal environment" (1996:55). Yet Binford and Kolata only hint at *why* these other researchers and development workers would represent the altiplano as a marginal environment. Binford and Kolata write:

> They [agronomists and development specialists] see a recent history of human underutilization of the altiplano and then assume that it stems from inherent environmental limitations. They have not yet appreciated that the more relevant determinants of underutilization are sociological, historical, and economic. Catastrophic demographic collapse, internal migration driven by national and international economic forces, loss of traditional cultural practices, a numbing history of racially based oppression—these are the more germane elements for explaining recent underutilization of the Andean altiplano. The environmental and agroecological image of the development specialists remains uniformed by long-term historical and cultural perspectives on human utilization and realization of altiplano ecosystems potentials (Binford and Kolata 1996:56).

Following on the above assertions that, though the lake basin may have been a difficult agricultural environment it was by no means marginal or inhospitable, one underlying theme throughout the collection of articles is that with appropriate social organization and technology this difficult environment can be mastered for intensive agricultural production. Thus the authors in the volume continually focus on the raised field agriculture as a lost indigenous technology of a former pre-Hispanic state. There is no explanation as to why the raised fields were no longer in cultivation, except to offer an environmental determinist explanation for why the raised field system collapsed during the reign of Tiwanaku (Erickson 1999). Therefore, according to the vision of the authors in this edited volume, the job of applied archaeology is to recover this lost technology, explore its function and practical application, and magnanimously bestow their gift of rediscovery onto the current indigenous populations of the Lake Titicaca Basin.

That the local lake basin farmers often did not appreciate this gift of a lost indigenous technology and recovered heritage was attributed to "cultural resistance" (Erickson and Brinkmeier 1991:14–16; Kolata et al 1996:209–10). As Kolata et al. write, "Overcoming the skepticism surrounding the feasibility of continuous cultivation, which stems from decades of experience with long-fallow systems, is difficult. With continuous encouragement from the project agronomists over the past five years, the number of farmers who accepted this dimension of raised-field cultivation practice has slowly increased" (1996:210). That the farmers may have resisted this farming method based on their own practical experiences and knowledge was downplayed. Most telling is the seeming disregard in all the publications for the amount of labor that the raised fields required. The expectation was that farmers could easily accommodate the demands for collective labor on this unproven agricultural technique.

The collections of articles found in *Tiwanaku and its Hinterland* represent raised fields in very scientific terms, forming the impression that the fields are sophisticated technological wonders that require considerable outside organization and management. Great detail is given to the engineering aspects of groundwater control (Ortloff 1996), the elaborate ecosystems produced by the raised fields, aqueducts, and causeways (Binford et al. 1996), and the nutrient fluxes and retention on raised fields (Carney et al. 1996). The raised fields were also represented as producing near miraculous functions for cultivating plants. For example, they describe the heat conduction of the fields which aided in preventing frost damage (Kolata and Ortloff 1996), how the nutrient recycling theoretically aided the long-term sustainability of production (Carney et al. 1996), and how the sophisticated hydraulic engineering of the fields' design drained surface water and maintained optimal water levels in the canals (Ortloff 1996). All of these images produce a representation of raised fields as a sophisticated and complex form of landscape management.

Of course, farmers and horticulturalists make raised mounds for planting the world over, from the elaborate pre-Hispanic *chinampas* of Mexico to simple raised beds in American backyard gardens. However, the representation of raised fields as sophisticated and complex in Kolata's (1996b) edited volume, implies that they are a complex form of agriculture that *required* state management and control. Further, this complex system of agriculture that the authors describe lends itself to the representation that the pre-Hispanic people of Tiwanaku were a complex, sophisticated, urban society. Perhaps not surprisingly, this depiction of pre-Hispanic raised fields also serves to support the view by Kolata (1993) that Tiwanaku was politically organized as a state, a hypothesis that other archaeologists still take issue with.

This is a rather tidy contrast to the contemporary Aymara-speaking farmers who live in the area and who are essentially represented in contrast as simple, rural, and backward farmers in Chapter 9 (Kolata et al. 1996). However, unlike development strategies prior to the 1980s that used modern science as the answer to the problems of poor agricultural production and poverty, the researchers propose that the answer to rural development lay in the secrets of the past, a past derived from a complex, urban, and sophisticated society at Tiwanaku (Kolata 1996b).

The representations of the past and the present that are being depicted in the articles by this team of archaeologists and other scientists have produced an equally essentializing and one-dimensional image of Tiwanaku, the Lake Titicaca Basin, and the contemporary people who live there. It is not new or original to maintain that archaeology, a field that interprets and reconstructs the lives of peoples in the past, produces images and representations of past peoples and societies that are interpreted through the lens of contemporary social-political relationships (Lowenthal 1990; Schmidt 1995). In representing and constructing the past, archaeologists are regularly involved in the politics and social relationships of the present. Nor is it a fresh observation that archaeologists have often been ambivalent to the impact they have on current populations. Though North American archaeologists are usually trained as four-field anthropologists, Schmidt and Patterson (1995) note that many still remain uninfluenced by wider anthropological perspectives. Therefore, many archaeologists remain uninfluenced by the reflexive analysis of cultural representations in ethnography characterized in the work of cultural anthropologists in the 1980s and 1990s, as illustrated in the beginning of this chapter.

Research has shown how archaeology has played a role in the politics of modern history by shaping ethnic consciousness and providing modern links to the past through the interpretations and cultural representations of archaeologists (Benavides 1999; Schmidt 1990). However, the question I wish to further explore is specifically how archaeology in Bolivia corresponded with other social and political relationships in Bolivia. Carlos Mamani Condori (1989) addresses this issue by arguing that archaeology in Bolivia has been a "legitimator of colonialism." Mamani Condori makes the link between archaeology in Bolivia and the triumph of the National Revolutionary Movement (MNR) in 1952. Aside from the early work of Wendell Bennett, systematic archaeological research at Tiwanaku and other northern Bolivian altiplano sites did not begin until the late 1950s (Kolata 1996a:6). The inception of archaeological research at Tiwanaku coincided with the formation of mestizo ideals that were being promoted by MNR following the 1952 national revolution. As Mamani Condori writes concerning this mestizo national project:

> Archaeology had an important role in this project it had the job of providing the new nation with pre-Spanish cultural roots. The object of their concern was to integrate pre-Spanish archaeological remains into the 'Bolivian' cultural heritage, and at the same time to integrate the Indian population into the stream of civilization (another of the main nationalist projects) (Mamani Condori 1989:47).

Mamani Condori does a thorough job of illustrating how early Bolivian archaeology was directly tied to mestizo nationalism in Bolivia. Tiwanaku is a special source of Bolivian national identity, and it is regularly depicted in Bolivia as a once powerful and expansive state whose epicenter is located on the Bolivian altiplano. More recently, archaeology in highland Bolivia led by North American archaeologists has contributed less to this Bolivian nationalist discourse, since it is less Tiwanaku-centric in its approach and focuses more on other cultural groups and pre- and post-Tiwanaku polities.

Yet I argue that the raised field rehabilitation project ultimately represented and came to symbolize mestizo nationalism in Bolivia, and not indigenous Aymara ethnicity. Though the archaeologists who were the early proponents of raised field rehabilitation may have had the best of intentions for restoring an indigenous technology for the benefit of rural indigenous peoples, as I will illustrate in the following section, the project quickly began to appropriate a nationalistic fervor and adopted discourses of development, specifically sustainable development. Like the discourses of *Katarismo* that was eventually absorbed into the multicultural symbolism of the state politics in the late 1980s and early 1990s, the raised field project also became a symbol of the pluri-national state in Bolivia. Though unlike the *Katarista* movement, which rose from within the ranks of indigenous society, raised fields is an invented tradition of indigenous knowledge by North American archaeologists and mestizo NGO workers in Bolivia. Ultimately, the raised field rehabilitation project failed to gain the full participation and long-term support of the inhabitants of the Bolivian Lake Titicaca Basin, the supposed descendants and inheritors of this indigenous technology.

It would seem that given such a recent and thorough self-reflection on the ways that anthropologists represent their subjects, archaeologists would have been extremely hesitant to uncritically apply a model of agriculture based on contemporary archaeological interpretations of the past. Rather, in the raised field rehabilitation project archaeologists became romantically and uncritically involved in promoting their own interpretations of the past though an internationally funded development project on the altiplano. This vision of the future proposed by archaeologists and development workers held that the answer for generations of agricultural-ists in the Lake Titicaca Basin lie in recovering a technology that had been lost and was now being rehabilitated through science and archaeology. In fact, such romantic visions of recreating a pre-Hispanic agricultural system seems, albeit in hindsight, to have been extremely naïve and entirely self-serving. Researchers involved in the raised field rehabilitation project did not take the time to reflect on the values and politics that shaped their own pro-environmentalist and pro-indigenous representations of Tiwanaku, the Lake Titicaca Basin, and contemporary peoples of this region.

REPRESENTATIONS OF RAISED FIELDS AND INDIGENOUS PEOPLES IN DEVELOPMENT

I now turn to the representations of indigenous peoples and technology in raised field agricultural development projects in the Lake Titicaca Basin. In the following section, I outline three themes in the representation of raised fields as ecologically sustainable, indigenous knowledge, and appropriate technology. These representations were found throughout the literature written and handed out to the communities that participated in the project, to funding agencies, and to the general public. These three themes in the representation or raised fields and indigenous peoples also emerged in informal interviews with various NGO workers and archaeologists who had participated in and promoted the Bolivian raised field rehabilitation project, as well as other attempts to rehabilitate raised fields in the Lake Titicaca Basin. To a lesser extent, some of these representations resurfaced in interviews with community members from Wankollo who had partici-pated in the raised field project. In the following sections, I depict how the raised fields were represented without necessarily trying to support or debunk the claims and assumptions associated with these representations.

Raised Fields as Ecologically Sustainable

The first theme is raised fields as ecologically sustainable. The very concept of sustainable development and sustainable agriculture has its roots in the conservation and environmental movements of the 1960s and 1970s (Adams 1990). By the 1980s, there was an emerging movement within the international development community towards environmentally friendly projects. This paradigm shift in development was called sustainable devel-opment. With the release of the Brundtland report in 1987, sustainable development took center stage in development discourses. Sustainable development discourses link economic development and environmental issues so that both topics became a single issue, inseparable from the other.

Sustainable development is defined in the Brundtland report as "development that meets the needs of the present without compromising the ability of future generations to meet their own needs" (Brundtland 1987:43).

One of the most consistent themes in the literature about raised fields and from interviews with persons promoting contemporary raised fields is that they are ecologically sustainable. Erickson writes:

> From a technical and social point of view, raised field agriculture of the Lake Titicaca Basin could be considered *a sustainable agricultural system* because of its high efficiency, low capital input, low maintenance, and high continuous productivity over the long run (Erickson 1992b:297).

Even when the raised fields are not outright labeled as sustainable (Erickson and Candler 1989), they were still portrayed in terms that are consistent with the image of sustainable agriculture. For example, the fields were promoted as not needing expensive chemical fertilizers because they created their own green manure in the muck at the bottom of the canals. Thus the fields were supposed to be self-fertilizing and "regenerative," another catch phrase of sustainable agriculture. Further, because they did not use chemical fertilizers and pesticides, some development personnel would also label the fields as organic agriculture.

For example, see Figure 8—Raised Fields as Organic Agriculture. This is an example from a pamphlet on raised field development handed out by the NGO. In this figure, it is encouraging farmers to fertilize their raised fields with organic dung fertilizer *("coloca tierra organica [huano]")*. While in Figure 9—NGO Pamphlet Cover 1, it shows the cover of another NGO Pamphlet that was titled "Raised Fields: Natural Horticulture." These same sentiments about the fields being "organic" and "all natural" were repeated to me in the community of Wankollo, where I conducted field research. However, I wonder where community members picked up these terms, and presume it was from the development project, since the terms "organic" and "all natural" were certainly not a part of the Aymara lexicon of agriculture.

Raised Fields as Indigenous Knowledge

The second theme is raised fields as indigenous knowledge. In the late 1980s and early 1990s, there were a number of indigenous social and political movements in Bolivia and across Latin America. In Latin America, indigenous movements were most visible leading up to the 1992 quincentennial marker of the landing of Columbus in the Americas (Hale 1994). In Bolivia, the Aymara *Katarista* movement brought indigenous discourse into the national dialogue during the 1980s, with indigenous marches and rallies becoming more frequent following the economic crises of the mid 1980s. By the early 1990s, political platforms and political discourses that celebrated indigenous consciousness and cultural diversity were a part of all the major political parties in Bolivia (Albó 1994, 1995; Rivera Cusicanqui 1993).

Likewise, an emphasis on indigenous knowledge and indigenous technology is directly linked to the sustainable development paradigm and was being ardently promoted by practicing anthropologists working in the development sector, led by the work of anthropologist Michael Warren (1991, 1999). Indigenous knowledge is local knowledge and is defined in contrast to Western scientific knowledge. According to Warren (1991), it is knowledge that is based on careful observation and understanding of a local environment and its natural resources. Recent critiques of indigenous knowledge question the distinction between fixed categories of indigenous and Western (Agrawal 1995). Such critiques recognize that all spheres of knowledge are embedded in specific political situations, which is the larger issue that should be addressed when privileging one form of knowledge above another, rather than polarizing the debate into opposing categories of indigenous vs. Western.

For example, in both in Figure 9—Pamphlet Cover 1 and in Figure 10—Pamphlet Cover 2, raised fields are represented as indigenous. In both of these examples from pamphlets produced by the NGO, raised fields are defined as a system of agriculture from the Andes. In Figure 9, it shows Aymara men in traditional clothes and playing traditional Andean instruments. While in Figure 10, it represents the raised fields by using pre-Hispanic Tiwanaku stylized iconography. The use of this Tiwanaku-esque iconography, such as from the cover of another pamphlet (see Figure 11—Pamphlet Cover 3), symbolizes the indigenous ancestry of the raised fields technology. While on the inside page of yet another pamphlet handed out to farmers (see Figure 6—Staff God), it describes the connection between Tiwanaku and the raised fields. In Figure 6, the NGO is symbolizing the indigenous heritage of the raised fields with a picture of the staff god found on the gateway of the sun at the site of Tiwanaku.

Raised fields and the raised field rehabilitation project were regularly represented as a form of indigenous knowledge. The difference with typical studies of indigenous knowledge is that raised fields are considered a "lost" indigenous knowledge, which through the research of archaeology was being recovered for use by the indigenous descendants of the pre-Hispanic Tiwanaku civilization. The NGO *Fundación Wiñaymarka* considered its mission as "the recovery of the Andean culture" and used stylized Tiwanaku iconography to symbolize the indigenous ancestry of raised fields. In interviews with the NGO workers, and the director, it was emphasized that they were rehabilitating a lost knowledge and giving it back to the descendants of Tiwanaku.

In academic works on raised fields as applied archaeology, there is the sense that raised fields are an indigenous cultural knowledge that is being rescued, recovered, and restored by archaeologists. In an article by Kolata et al. (1996) the authors write that "agroecology," or sustainable agriculture, uses an ecological systems approach to search for local (read indigenous) agricultural alternatives. Kolata et al. (1996) define this agroecological approach as an attempt to understand agriculture in its local cultural context vs. what they call an "imposed model of industrial agriculture." The authors have linked their ecological approach to alternative agriculture, and to the study of local or indigenous systems of agriculture, in this case the study of raised fields. Later in the same article the authors write, "It is the resolution of the social and cultural problems of introducing raised-field agriculture to rural communities that will be critical to its long term sustainability" (1996:229). Yet I cannot help wondering that if the raised fields really are an indigenous knowledge and a local alternative to imposed Western development, why do Kolata et al. (1996) and the development NGO say that they face problems introducing it into local communities? If raised fields really are an indigenous and local knowledge derived from the local or indigenous cultural system, why must they now be reintroduced back into the lake basin?

Raised Fields as Appropriate Technology

The third theme in the representation of raised fields concerns the fields as appropriate technology. The concept of appropriate technology is linked to both of the previous two themes. Technology, in this case the raised fields, are considered appropriate because the are not capital intensive, a resource in short supply for subsistence agriculture, and are easily available to local farmers. Another example of this theme from Erickson reads:

> A more effective approach to development is through what is referred to as *"appropriate technology."* This approach stresses the use of traditional forms of technology and ecologically sound modern forms that are not capital intensive. In the Andes, there is a large work force available, but little capital. Since communal work forces are the traditional form of labor organization, an appropriate technology that is easily adopted by peasant communities would involve cooperative labor (Erickson 1992b:15).

In this passage, Erickson maintains that raised fields are appropriate because, rather than investing large sums of capital, raised fields are built by human labor organized in "traditional" work groups and using simple indigenous tools. One of his assumptions is that there actually is a large work force available for subsistence agriculture.

However, the use of the term appropriate technology is rather absurd, since the factors that determine whether a technology is deemed appropriate for local development is based on political economic factors instituted by hundreds of years of oppression and exploitation. The argument goes that because a farmer is poor, he or she should only use manual labor since such technology is appropriate to his or her economic means. The fact that a farmer is disadvantaged because of hundreds of year economic development by the West seems to be disregarded in the appropriate technology discourse. This leads into something of a self-fulfilling scenario that serves to maintain inequality rather than to help alleviate it.

The rehabilitated raised fields were represented and promoted as appropriate technology by emphasizing that they are built without machinery or tractors, and using only indigenous hand tools for cultivation. In pamphlets and published articles on raised fields, it is emphasized that participants only use indigenous tools and manual labor. For example, see Figure 12—Tools for Raised Field Cultivation, which shows the tools depicted in one pamphlet that a farmer might use for raised field cultivation. In another example from Kolata et al. (1996:206), it shows a drawing of "traditional Aymara hand-held farming implements." These particular "traditional Aymara" tools rather looked like something an archaeologists might dig up—and somewhat prehistoric—while I never actually saw any of them being used for agriculture in Wankollo in 1996–97. In fact, I never actually saw any of them in the homes and compounds of Wankollo residents in my rounds of interviews during the year that I lived and worked with them. Instead, when manual labor had to be done, Wankollo residents used metal picks and shovels, while preferring the use of oxen plows and tractors to cultivate their agricultural fields. The NGO actually distributed metal picks and shovels to local farmers as an incentive for participation in the project, though these tools are rarely shown in the images of raised fields in pamphlets and articles.

DISCUSSION

Throughout the representations of raised fields, the three themes of the fields as ecologically sustainable, indigenous knowledge, and appropriate technology resurface again and again. Symbols of the pre-Hispanic Tiwanaku civilization are found throughout the publications of the NGO *Fundación Wiñaymarka*. This is not surprising since the Bolivian development personnel at the NGO consider their development work to be the recovery of an ancient technology. This is noted in the title of the NGO, *Fundación Wiñaymarka*, which in Aymara means eternal or timeless nation. The subtitle of the foundation is "for the recovery of the Andean Culture" *(para el rescate de la Cultura Andina*—see Figure 9 and Figure 10), which shows a conscious connection being made between the development project and the ancient Andean civilization. Another example of Tiwanaku iconography depicted in representations from the development group includes an image of an Andean cross that is found on the inside back cover of another pamphlet (see Figure 5—Andean Cross). This depiction of the cross is based on images found carved into monoliths at the site of Tiwanaku.

The themes of ecologically sustainable agriculture, indigenous knowledge, and appropriate technology can be found in the following passage from a final report written by the NGO *Fundación Wiñaymarka*:

> The investigation and reevaluation of the pre-Columbian agricultural technology, demonstrates that these are not the result of a theoretical formula, if there is not close contact with Mother Nature, an understanding of the laws and principles that govern the universe, and respect for life and the admiration for all of creation. In conclusion, the ancestral agricultural production systems did not seek out the exploitation of natural resources for one's own advantage with the objective of generating profits, instead they were directed at promoting harmony with the environment and the well-being of the community as a whole, making rational and sustainable use of that which the *Pachamama* [mother earth spirit] would offer to man from her bosom, without depleting or pillaging the environment. This cosmic conscience existed in the past, of which all were part of the universe and like such they could not destroy the only house that they had, the planet Earth. Therefore, their efforts were destined to preserve the immense wealth that was given to them, instead of depleting it and being dragged in the catastrophic fall of the degradation of the environment, that means the death of all of creation.
>
> For these reasons, the ancient inhabitants of the altiplano knew how to face the adversity of the environment, creating, thanks to his intelligence and conscience, production systems that were in harmony with nature. They took advantage of, to the maximum advantage that was offered, ways to prevent and to diminish the risks and the losses by opposing events, such as the frosts, the droughts and the strong winds.
>
> From patient observation of nature, as well as the benefits that could be obtained by means of the fitting together of their rudiments and resources, is born *Sukakollus* [raised fields], which, as it is understood today, are a production systems in which they obtain the maximum biotic and abiotic performance of components, for the benefits of the human community (Fundación Wiñaymarka 1995:8–9).

In the above passage, we can see many elements from the themes of ecological sustainability, indigenous knowledge, and appropriate technology. For example, there is the emphasis on the ecological sustainability of raised fields through the multiple references to the "sustainable use" of natural resources, and the "preservation" of the environment. The repeated reference to the fields being "in harmony with nature" also implies that they are ecologically compatible to the lake basin. Further, the raised fields are sustainable because the production system does not "deplete" or "pillage" the environment for the sake of profits, instead preserving the earth's bounty for all of humankind.

The passage also has multiple references to the "pre-Columbian" and "ancestral" pedigree of the fields, making the link between the contemporary raised field project and its indigenous past. There are repeated allusions to "mother nature" and the *Pachamama* (Andean mother earth spirit), both suggesting that this ancient technology was "natural" and has mystical ties to the Andes and the natural environment of the Lake Titicaca Basin. The production system is also represented as appropriate technology, since it is described as using only rudimentary technology and the natural resources of the lake basin that was handed down from the "ancient inhabitants of the altiplano."

Visual representations in numerous development publications also depict Aymara speaking farmers wearing "traditional" clothing. By traditional, this means typically non-European clothing that is associated with rural peoples, and including the Andean ponchos. Women are shown wearing the full *pollera* peasant skirts, *mantillas* (shawls), and bowler hats. This indigenous style of dress is a marker of both ethnicity and class in Bolivia. Changing one's clothing to Western styles can indicate a change, or an attempted change, of economic class. For example, people changing from an agricultural lifestyle to urban working class or middle class would likely adopt a more Western style (Harris 1995).

Farmers are also shown using traditional Andean tools, as well as other manual tools for building and cultivating agricultural fields (see Figure 12—Tools for Raised Field Cultivation). Rather than suggest that farmers do not use these types of tools, I want to place emphasis on the messages embedded in promoting the use of so-called indigenous, traditional, and manual technology. By focusing on the use of traditional tools, particularly the Andean footplow, these images highlight that the farmers are indigenous to the Lake Titicaca Basin and are of pre-Hispanic origin. It emphasizes and reestablishes that they are the inheritors of the Tiwanaku civilization. The use of traditional tools, and the focus on manual labor, shows a certain sense of these peoples as pre-modern and not dependent on modern agricultural technologies.

Field research in Wankollo revealed that most families preferred to hire a tractor to turnover new soils for cultivation and used oxen for further working of the soils and planting. As stated above, I never once saw an Andean footplow in the Tiahuanaco Valley during my two years in the field. I have only seen them depicted in the development pamphlets and have never actually seen any of them in farmers' homes. Rather, farmers in Wankollo were interested in decreasing the amount of time spent cultivating fields so they could divert more labor into off-farm economic activities.[2]

Throughout the development representations it is emphasized that community members should work together. There are systematic representations of community members working together in group work parties. For example, see Figure 13—Representation of Communal Work. In this example from a NGO pamphlet, it urges community members to work together saying "The work is easier if it is shared." The idea that this type of group work party is a traditional way to organize labor in the Andes is a common theme throughout the publications and pamphlets on raised field rehabilitation. For example, Erickson describes how his team of researchers working in Peru proposed that each community organize work in the "tradition of *mink'a*" (festive labor party), with each family sending one adult to work during the community workdays. However, Erickson does note that this type of group work party, which he calls a "traditional" form a labor, was eventually replaced by the participants themselves who preferred a method of labor organization called the "*tarea*" (task) system. The *tarea* system is where each household is assigned a specific plot of land to build and cultivate, requiring that they organize their own labor for the task (Hirst 1998a).

During fieldwork in Wankollo in 1996–97, I saw few communal work parties, and those work parties that were organized were for projects such as repairing roads and community buildings. This practice of forming a community wide work party was not used in agriculture in the season of 1996–97. All the raised field project leaders complained to me that organizing and motivating workers to participate in the community level agricultural project was difficult. For agricultural work, often groups of 2, 3, and up to 5 persons gathered for planting or harvesting, but this was generally the limit.

Many of the images and representations of indigenous peoples depicted in the pamphlets and publications on raised fields look remarkably similar to images produced by the 16th century Andean chronicler Felipe Guaman Poma (see Figure 14 and Figure 15—Guaman Poma 1 and 2). Considering the era in which Guaman Poma produced his images, they also emphasized indigenous peoples using traditional tools, and often showed them laboring in work parties. Of course, these images are very familiar to scholars of the Andes, and probably to the literate upper and middle class mestizos in Bolivia as well. The similarity between the images in the development pamphlets and the 16ᵗʰ century images of Guaman Poma draws an additional link between the project and its supposed ancestral and indigenous roots. The development project wanted to recover a lost technology for use by current-day farmers. In a publication by the NGO *Fundación Wiñaymarka* on the rebuilding of pre-Hispanic terrace agriculture, the NGO actually used these same 16ᵗʰ century images from Guaman Poma as illustrations for agricultural production for current 20ᵗʰ century farmers.

CONCLUSIONS

In conclusion, the NGO *Fundación Wiñaymarka* represented the raised fields as indigenous and traditional, while emphasizing that the raised fields were an ancient technology that could help the present day farmers of the Lake Titicaca Basin. They used ancient Tiwanaku iconography, illustrations of farmers in traditional clothes and using traditional manual farming tools, and portrayed the farmers working in traditional group work parties. Following Hobsbawm (1983), I argue that the invented tradition of raised field agriculture emphasized manual technology, while corresponding with sustainable development concerns for "appropriate technology." Emphasis in the raised field development project was placed on the manual building of fields that relied on human labor and not modern machinery. According to archaeologists and the NGO, all it took to build the field was a pick or a shovel, not expensive machinery and tractors. Of course, the assumption underlying this emphasis is that farmers in the Lake Titicaca Basin had plenty of labor available to build and maintain the raised fields. As I will show in chapter 6, this was patently untrue.

The fields were also promoted as all-natural and self-fertilizing, harkening back to the conservationist roots of the sustainable development movement. The archaeologists themselves promoted the fields as self-fertilizing by the creation of green manure, which purportedly carried enough nutrients to replenish fields for annually production of potatoes without fallow. Both Erickson and Brinkmeier (1991) and Kolata et al. (1996) cite cultural barriers in trying to persuade Lake Titicaca farmers to replant raised fields in potatoes after 1 to 2 years of cultivation, yet neither has produced the long-term data that suggests that annual planting of potatoes is feasible. While practical experience in Wankollo, and across the Tiwanaku and Catari Valleys, demonstrates that raised fields cannot be replanted annually in potatoes.

The promotion of raised fields was a mestizo appropriation of Tiwanaku as a symbol of the modern nation-state of Bolivia, which created continuity between the modern mestizo nation and the pre-Hispanic past. The raised field development project was an attempt to create an invented tradition by literally rebuilding the ancient agricultural system that once supported a pre-Hispanic civilization. Such imagery creates a link between the modern, struggling democratic nation of Bolivia and the expansive, urban, and sophisticated pre-Hispanic civilization of Tiwanaku. This alleged expansive pre-Hispanic state is strikingly illustrated on an inside cover page of one pamphlet on raised field development that shows the area distribution of this ancient agricultural system placed within the contexts of an expansive Tiwanaku State (see Figure 7—The Tiwanaku State). This map delineates a Tiwanaku State whose empire encompasses all of highland Bolivia and a long stretch of coastal lands now belonging to Chile and Peru. Whether Tiwanaku was a state, and actually had direct control of these vast coastal areas, is still contested by archaeologists. However, where control of these areas by Tiwanaku did exist, it certainly did not resemble the way modern nation-states control and govern peoples and land areas. Yet the sentiments that these lands once "belonged" to Bolivia, and that they were lost in successive wars after gaining her independence, is a common theme in Bolivian nationalism. Bolivian demands for direct access to coastal ports is still an item on the agenda of Bolivian international politics. Yet this map found in a raised field rehabilitation pamphlet shows the ancient and vast Tiwanaku state as encompassing large stretches of coastal land now under Chilean and Peruvian rule. What the map is representing is 20th century Bolivian nationalism that is being constructed and revitalized through the recovery of this pre-Hispanic technology through the raised field rehabilitation project.

The irony of Bolivian nationalism that draws on Tiwanaku as a symbol of a historic and expansive state is that it is the non-Aymara people of European ancestry, the ones who have oppressed the indigenous peoples for 500 years, who have appropriated the indigenous symbol of Tiwanaku and are promoting this indigenous knowledge of raised field agriculture. The contemporary Aymara residents of Wankollo expressed little interest in recovering this lost technology of raised fields. When the food and tool incentives were discontinued, and the raised fields stopped producing potato crops, the residents returned them to fallow and did not build any new raised fields.

But what did the project succeed in doing? The raised field project created an image of timeless and traditional Aymara farmers that *attributed* a collective and indigenous identity onto a subaltern rural population (Comaroff 1987:305). Invented traditions like the raised field project imply certain acceptable social behavior and social markers, in this case that the farmers are indigenous, rural, and traditional peasants or subsistence based agriculturalists. This is counter to the nationalistic ideas that the NGO workers had of themselves, as modern, urban, and mestizo Bolivians. The portrayal of indigenous peoples in traditional garb and using traditional farming methods, in effect, maintains the class and ethnic boundaries between them, and the urban, mestizo, middle and upper classes. When indigenous knowledge and indigenous cultural symbols are appropriated and reproduced by a dominant group, even if the intentions are altruistic, the results are likely to fall short of development goals.

NOTES

[1] Agricultural production on raised fields is examined in more detail in chapters 5 and 6.

[2] See chapter 6 for an extended discussion of labor availability in Wankollo.

CHAPTER FIVE

Traditional Agriculture Practices: Contrasting Representations of Raised Fields with Production Factors at the Local Level

W ITH ALL OF THE INITIAL ENTHUSIASM BY ARCHAEOLOGISTS AND other researchers, the coverage by the press, and subsequent funding by various international development agencies, it seemed that raised fields offered a viable solution from the past for contemporary agricultural development in the Lake Titicaca Basin. Yet by 1994 most of the raised fields built by the Bolivian raised field rehabilitation project were already beginning to be abandoned across the altiplano and no new fields were being built. After 2 to 4 years of cultivation, most of the communities that participated in the project stopped building new fields and discontinued cultivation on fields that had already been built. Yet community members in Wankollo did not think of the project as being a failure. For community members who had participated in the fields it had been a worthwhile venture. They had had several years of good potato production on the fields, which had provided additional food and crop seed. The also received various tools, potato seed, and foodstuffs as incentives for participating in the project. Certainly in the eyes of the community, the project was not a failure, since they had participated in the project, reaped some significant short-term rewards, and continued with their normal dry field farming techniques without any ill affects in the community. In the community's view the project had provided increased subsistence goods, though only for the short-term.

However, based on the development group's objectives and on the predictions of the archaeologists, the raised field rehabilitation project did not live up to expectations. First, development groups and archaeologists had envisioned the raised field rehabilitation project as a practical project that would rebuild the raised fields and this ancient agricultural system, and also as an educational enterprise that would teach the farmers to continue building raised fields on their own. One long-term goal of the rehabilitation project was for the eventual withdrawal of development aid and the continuation of raised field rehabilitation and cultivation without assistance from the NGO. Development leaders imagined that once taught this remarkable ancient method of raised field cultivation, which required only crop seed, access to appropriate land, and copious human labor, farmers would spontaneously continue to build and cultivate more raised fields without project leadership. Yet when development incentives, project leadership, and the organizing of work groups were discontinued, so too was the building and rehabilitating of new raised fields.

A second indicator that the fields had not lived up to development predictions and expectations was that none of the fields in the Bolivian Lake Titicaca Basin were planted for more than 2–4 years before being returned to fallow. Most early researchers had proposed that the fields were sustainable and regenerative, allowing farmers to continuously cultivate the fields in potatoes each year. Yet farmers across the Catari and Tiahuanaco Valleys discontinued cultivation on individual fields after 2–4 seasons.

In chapters 2 and 3, I recounted the long-term political and economic strategies of the Aymara in the Lake Titicaca Basin by examining the social and economic history of the region. I also looked closely at more recent trends in Bolivian economy development policies, indigenous social movements in La Paz, and in international development in order to consider the immediate contexts for the raised field project. In chapter 4, I outlined the history and representation of raised fields as depicted by archaeologists working in the Lake Titicaca Basin and by NGO workers who reproduced these representations when they devised and implemented the raised field rehabilitation project. It was the NGO and the development project that put these collective representations of the fields, and of the people who would build and cultivate the fields, into contemporary practice. The primary research problem I explore in the following two chapters is the differences between how the fields were represented by researchers and the development NGO, and the economics of raised field cultivation in practice in contemporary communities in the Bolivian lake basin. I also integrate local views on raised field cultivation and farming in the Lake Titicaca Basin by community members from Wankollo.

In this and the following chapter, I take a close look at the economics and practice of raised field cultivation drawing on several sources: interviews with farmers, NGO workers, and archaeologists; oral histories of raised field cultivation with participants from Wankollo; data gathered on farming practices and rural economy in Wankollo; and published reports and unpublished manuscripts about production on raised fields by researchers and development groups.

In this chapter, I break down my discussion into four themes about production and cultivation on raised fields. The first theme that I discuss is the solar heat retention and microclimate effect produced by the water in the canals surrounding the fields and the amount of protection from frost that this provided to crops. The second theme concerns the regenerative aspect of the fields, which was supposed to allow for continuous cropping of the fields in potatoes and provided the ecologically sustainable image of the fields. The third theme regards the use of appropriate technology, specifically the use of hand tools for building, cultivating, and maintaining the fields, which lent the fields the image that they were economically appropriate to the smallholder farmers and economically sustainable. The forth theme that I explore is access to land for raised fields, where I address the question of whether land was a significant factor hampering the cultivation of additional fields by households in Wankollo.

All of these production factors lead to what I consider the primary obstacle for contemporary cultivation on raised fields: access to agricultural labor and the social organization of agricultural labor within the contexts of current Bolivian political economy. The consideration of the role of labor in agriculture requires a chapter of its own to address the many complex and inter-related issues falling under this topic and will be addressed in chapter seven.

AGRICULTURAL PRODUCTION IN WANKOLLO

Agricultural production in Wankollo, and the Lake Titicaca Basin, is organized around the cultivation of potatoes. The elevation of Lake Titicaca and the southern lake basin region is about 3,800 meters (about 12,500 feet) above sea level. Since maize cannot be grown at the high altitude of the altiplano (except on very favorable soils in protected areas next to the lake), potatoes are the staple crop for the local population.

There are two principle types of potatoes that are grown: sweet and bitter. Sweet varieties of potatoes tend to be larger and are quite susceptible to freezing temperatures. These are the varieties that North Americans are familiar with, and they are usually boiled or steamed for consumption. The bitter varieties are much more frost resistant and are very well adapted to the high altiplano. The most common variety used in Wankollo was the bitter variety, *"luki."* The bitter varieties are usually consumed in their dehydrated forms as *chuño* or *t'unta*. There is no market for the bitter vari-

eties of potatoes, and I know of no families who regularly sold any potatoes in the market.

Other cultivated crops that are grown include the native *quinoa* and
cañahua (two types of chenopods), fava beans, and the root crops *papa
lisa, oca,* and *tarwi.* A number of introduced crops are also grown for
human and animal consumption, including barley and oats, and a few
households have gardens with onions and other herbs. In Wankollo, there
is a community greenhouse attached to the Natural Hospital that grows
lettuce, onions, tomatoes, and peppers, which are distributed to community members at the monthly meetings when the crops are in season.

There is only one planting season for potatoes on the high plains of
the altiplano, and this season is very short in comparison to production at
lower altitudes in the Andes. In Wankollo, planting begins after the advent
of the first seasonal rains. There are basically two seasons on the high
plains: the cold, dry winter of April through September, and the much wetter and milder period of October through March. Precipitation is heaviest
in December through March, with little or no precipitation in the months
of May to the end of August. Precipitation varies widely on a yearly basis
as measured at El Alto on the altiplano. From 1946 to 1992, average annual precipitation was 589 mm, with the maximum recorded for one year at
870 mm and a minimum of 325 mm (Crespo 1993:124).

The tools that farmers used or had access to for cultivating their
fields and processing their crops in 1996–97, included tractors for the initial plowing of new potato fields, Mediterranean style oxen drawn plows
for preparing, planting, and harvesting fields (Carter 1964), and numerous
manual tools. Most households hired someone to plow their new potato
fields by tractor when they were preparing them for the first potato crop
after a long period of fallow. Wankollo households hired either a local
cooperatively owned tractor, or another Wankollo resident who owned his
own tractor, to come and plow their fields. They paid for the tractors in
cash with the pay rate usually based on the size of the field that needed
plowing. After the initial plowing, further preparation of fields was done
with oxen drawn plows. These plows consist of a single draught plow with
a backward extending handle (Carter 1964). The plow is attached to a pair
of oxen by a yoke and is usually fitted with an iron blade. Other tools used
in cultivation include iron bladed picks, shovels, hoes, and sickles.

No chemical fertilizers were used on potatoes fields in 1996–97.
However, animal dung was carefully gathered over the course of the year
and used for fuel and fertilizer. Before planting a field in potatoes, farmers
usually transport several bags of sheep dung to their fields, which they

sprinkle into each furrow before dropping in the potato seed and covering the furrow.

The typical crop rotation on fields is one year of potatoes, followed by one crop of *quinoa* or fava beans, followed by 1 or more years cultivating barley or wheat before the field reverts back into fallow for a rest *(descanso)*. Most fields are cultivated for three years before returning to fallow (Carter 1964), though in a few select areas of Wankollo farmers may choose to plant two years of potato crops, followed by multiple years of barley or wheat. While a field is resting in fallow, it gradually becomes covered by natural vegetation and *ichhu* grass, and the field is used for animal pastures until the next agricultural rotation cycle. The length of the cycle and the fallow period depends both on the fertility of the soil, and on the pressure for land and land-scarcity.

The total land area of the community of Wankollo is 1,931 hectares, with all of the cultivable land in the community rain fed. The community does not have access to any high altitude fields for grazing animals. Most of the community land (94%) could potentially be cultivated, though some areas were more prone to seasonal inundation and tended to be left in permanent pastures. The fact that most Wankollo residents did not have access to nearby hillside fields for cultivation is an important issue, since these fields offer some protection against the periodic frosts that regularly plague the region.

Categories of land in the ex-hacienda of Wankollo follow a similar pattern as that found in free Aymara communities (communities that had never been owned by a Spanish *hacendado*) prior to the agrarian reform (Carter 1964). They are divided into two major types of land: *sayañas* and *aynokas*. The *sayañas* are usually identified as the household plots and consist of the house compound and lands associated directly with it. However, as Carter notes (1964), the *sayañas* are much more than a household plot and may in fact be quite large. For example, in the free community of Irpa Chico the largest *sayaña* was 48 hectares (Carter 1964:65), while in Wankollo the largest *sayaña* was 40 hectares. The *sayaña* has the most secure tenure and follows rules of partible inheritance in Wankollo. The second category of land is the *aynoka,* which is a large extension of land that is "owned" by the community. Though these are not private lands *per se*, access to specific plots in the community *aynoka* in Wankollo is inherited much the same way that access to *sayaña* lands are.

Areas of the altiplano floor that are regularly inundated during the rainy season are usually not cultivated and remain in permanent pastures. These are also the same areas that are most suitable for raised field cultivation due to their access to water. In Wankollo, seasonally inundated areas include the low fluvial area along the river drainage, and random pockets of low-lying depressions in the altiplano floor that fill with water and form annual temporary ponds during the rainy season. As I will discuss in the following section, these are seasonally inundated areas that were usually used as permanent pastures prior to the introduction of raised field cultivation in Wankollo.

Aside from the household plots *(sayanas)*, very few households in Wankollo had access to hillside fields (see Figure 4—Agrarian Reform Map of Wankollo). Unlike most communities in the region, the bulk of Wankollo's land consisted of the rolling floor of the altiplano. This is considered a disadvantage by members of the community, who value access to hillside fields for protection from the periodic frosts that plague the Lake Titicaca Basin. A household with access to hillside fields in Wankollo is considered fortunate, because these fields protect against frost, which drains down from the hillside and collects in cold pockets on the altiplano floor. Therefore, those areas of the altiplano that are seasonally inundated by water also carry an extra threat of frost.

However, there is one other factor that mitigates frost damage and offers crops some protection: the natural fertility of the soils. In areas of Wankollo deemed to be more fertile, such as the community held *aynoka*, farmers report increased frost protection and a general increase in plant health. It is these particularly fertile areas of the community that may be planted in potatoes for a second season, followed by several seasons of wheat or barley.

Finally, it is worth mentioning that most households in Wankollo did not produce enough crops for their own consumption. When planting began in September and October, most households used what potatoes they had left for seed and often had to buy more crop seed. After the planting and until the first harvests in April, most families in Wankollo had to purchase their potatoes for consumption. This demonstrates that households in Wankollo were dependent on off-farm income, since little or no income was generated through agricultural production.

THE RAISED FIELD REHABILITATION PROJECT IN WANKOLLO

The primary setting for research on agricultural production, land tenure, and labor deployment for raised field agriculture was the community of Wankollo, a community that had participated in the rehabilitation project under the promotion and guidance of the NGO *Fundación Wiñaymarka*. In 1989, one small ½ hectare plot of experimental raised fields was built in the community. The landowner of the field organized the labor for building and cultivating the field with the help of the NGO *Fundación Wiñaymarka*. In cooperation with the development project, Wankollo residents constructed approximately 2½ hectares of "community" raised fields in the following agricultural seasons of 1990–91 and 1991–92. I call these raised fields the community fields, because they were built on communal land and where worked by a large group of people drawn from all sectors of the community. In addition to the community raised fields, 3 individual landowners called upon their families and friends to build additional raised fields between ¼ to ½ hectare in size. These "private" family level fields were constructed with varying amounts of aid and consultation from the NGO *Fundación Wiñaymarka*. I call these segments of raised fields, private fields, because they were built on a single landowner's household plot (*sayana*), which has the most secure tenure and usufruct was similar to that of private ownership. Labor for the private raised fields was the sole responsibility of the landowner, who usually drew on extended kin and neighbors to built and cultivate the fields. In return, he distributed the crops produced on the fields equally among the participants, though he kept a larger portion for himself.

In 1996, the community of Wankollo consisted of 96 households and had approximately 400 residents. Most of the full-time residents of the community participated in the first year of the raised field project and helped to build and cultivate the community raised fields in the agricultural season of 1990–1991. In interviews, Wankollo residents said that they had been interested in building raised fields to take advantage of the enhanced and continuous production yields, and they were also interested in the frost protection benefit that the fields offered.

Although I could not obtain an official count, community members estimated that between 60 to 80 people helped to build, plant, and harvest the raised fields in the first year of cultivation on the community fields. This number dropped in the second year, with steadily decreasing numbers of participants until all the raised fields were abandoned throughout the community by 1995. The smaller plots of private raised fields built by individual landowners usually had between 10 to 20 people involved in construction and cultivation.

SOLAR HEAT RETENTION AND FROST PROTECTION

Frost is probably the primary agricultural threat to farmers cultivating potatoes and other crops on the high plains of the Bolivian altiplano, and Wankollo was no exception. Farmers in Wankollo considered themselves to have less favorable access to land, relative to other communities in the region, because the community consisted of primarily the flat plains (*pam-*

pas) of the lake basin floor with few hillside fields to cultivate. Other local communities in the Tiahuanco and Catari Valleys had more land diversity because residents usually have access to both the flat *pampas* lands and access to land in the nearby hills. Access to these higher altitude plots is important for the frost protection that these fields offer to crops, as well as for additional pastures for animals.

According to Wankollo farmers, the two areas that mitigate frost the best in the community are the very small number of hillside fields scattered throughout the community, and second, the communal fields *(aynoka)* that were located towards the far eastern most tip of the community. In the very low-lying areas on which the raised fields were typically built, frost is the primary threat to crops (seconded by flooding during particularly wet seasons). Hillside fields, on the other hand, offered some protection from frost since cold air masses collected primarily in low spots on the valley floor. The slope of the fields acts to drain cold air masses down the hillside and away from field surfaces. The problem with hillside fields is that they are much more prone to erosion, a factor that development agents seem to emphasize much more so than the community residents who value these fields for the frost protection they provide.

The second area that was reported as offering protection from killing frosts was the communal field called the *aynoka*. According to residents of Wankollo, fields planted in this area typically fared better during killing frosts because they had very good, nutrient rich soils. According to the farmers, the quality of the soil is a very important factor in how well crops produce even in the face of extreme frost. In areas with good soils, potato production was considered better than average and crops resisted frost better than in areas where the soils were considered poor. This factor has not been thoroughly examined or considered in evaluating the frost protection benefits of raised fields. For example, since the raised fields were usually built on land that had previously been uncultivated for generations if not centuries, the nutrient level in the soils for the raised fields would be very high. Therefore, if plants derive some frost protection benefits from soil quality and general plant health, then plant resistance to frost on newly planted raised fields would be elevated as well.

In order to understand the magnitude that the problem of frost causes on the high plains of the Bolivian altiplano, one cannot think in terms of merely a single night of frost. As a key informant described to me, a very bad year is when there are severe frosts for several nights in a row, even up to a week. This is the kind of frost problem that will kill entire crops throughout the community, and in neighboring communities throughout the valley. During the 1996–97 agricultural season there was only one single night of frost. This frost was described as mild and the 1996–97 agricultural season was considered a very good year for agricultural production in Wankollo.

Community members indicated that they were very interested in the frost mitigating capacities of the raised fields as well as the increased production yields. The primary reason given by Wankollo residents for participating in the community raised field project was to benefit from the frost protection that the raised fields were supposed to offer. Three farmers that I talked with described the situation during the first year that the raised field development was proposed in the community in 1990. These farmers stated that in previous years they had had several very devastating frosts that had damaged or completely destroyed crops in their community. One informant joked that many families were only eating rice and *yuca* (manioc) that year, crops that obviously do not grow on the high plains, thus implying that many families had lost their entire crop of potatoes due to the extreme frosts. Clearly, the reported frost mitigating properties of the raised fields had captured the interests of the community, particularly some of the community's younger leaders.

Though other researchers had written about the possible heat storage capabilities and frost protection that raised fields gave plants in this very frost prone region (Denevan and Turner 1974; Smith et al. 1968), Kolata and Ortloff (1989, 1996) came up with the first formal mathematical model of how this process of heat conduction and retention would work on the Lake Titicaca Basin raised fields. Simply put, the water that was maintained in the canals that surrounded the raised planting surface collected solar heat from the strong tropical sun during the day, which was released at night in the form of warm humid air. Recent research by Diego Sánchez de Lozada (1996) confirms this general hypothesis. By measuring a series of raised fields of various widths, lengths, and canal depths, and comparing air and soil temperatures with adjacent control plots of normal dry agricultural fields, Sánchez de Lozada found that differences in air temperatures between raised fields and control plots were confirmed. On cold mornings when killing frosts were most likely to hit, minimum temperatures were between 0.5 to 1.8 degrees warmer relative to the control plots with no raised fields (Sánchez de Lozada 1986:82). This slight difference in air temperature could definitely be the difference between total crop failure and only minimal frost damages during a mid-season frost. However, it certainly dispels some of the more extravagant estimates, which postulated that the heat retention aspect of raised fields might allow for double cropping of fields during the winter months when average nighttime temperatures typically fall several degrees below freezing from June through August (Kolata and Ortloff 1996).

Though the fields apparently do offer some protection to plants against frost, long-term tests of the heat storage properties have yet to be published if in fact such research has been completed. Meanwhile, anecdotal testimonials about the frost protection offered by raised fields in the first year of cultivation that aided in the miraculous survival of plants during a killing frost are not enough to confirm the extent and long-term protection that the fields offer. As Kolata and Ortloff write:

> Striking experimental confirmation of the heat conservation effects described here was obtained during the 1987–88 growing season in the Lakaya sector of the Pampa Koani. . . . On the nights of February 28 and 29, 1988, the Bolivian altiplano in the Pampa Koani region suffered a killing frost. . . . Many traditional potato fields within a few hundred meters of the experimental plots experienced crop losses as high as 70–90 percent. In contrast, losses in the experimental raised field of Lakaya I were limited to superficial frost lesions . . . (Kolata and Ortloff 1996:133).

This anecdotal evidence leaves much to the imagination and speculation of the reader by focusing solely on the high-end estimates of crop losses on nearby fields (rather than relating the average net losses to farmers) and neglecting to factor in the higher quality of soil on the previously uncultivated lands of the new raised fields. Since raised fields were built on previously uncultivated fields, while the comparison fields were on lands that had probably had only minimal durations of fallow in recent generations, these anecdotal insights are hardly a fair comparison. The comparison of fields built on land that had lain in fallow for decades (perhaps centuries), with fields that had been in continuous rotation with only short fallow periods, does not clarify whether the dominant frost protection benefit was derived more from the cultivation of a nutrient rich plot of land or more from the heat conduction from the canals. In any case, it certainly cannot discern the differential influence of each of these factors on the seemingly miraculous plant survival rates on raised fields in their first year of cultivation.

Wankollo residents who participated in the raised field project claimed that the fields did not protect the crops from frost as well as they hoped and expected they would. Like their regular dry potato fields, the primary problem for production on the raised fields was frost. The farmers' estimates ranged from the belief that raised fields offered little or no frost protection to the fields, to only slightly better than average protection.

As one informant explained to me, there was always a threat of frost in Wankollo, and this problem with frost affected the raised fields as it did the regular dry fields. Unfortunately, by the time of my own fieldwork in the Lake Titicaca Basin in 1996–97, none of the raised fields built by the development group were still in cultivation on the altiplano. Therefore I rely on oral histories for information on raised field practices.

According to Kolata et al. (1996: 217–28, 230), the raised fields in the community of Wankollo were not managed as well as recommended by the development group and at times certain sections of the fields did not maintain adequate water levels. They write that production levels on raised fields in Wankollo were lower relative to other communities due "to internal organization problems and incomplete grasp of the techniques entailed in raised-field construction and maintenance" (Kolata et al. 1996:217). The authors claim that the water levels on Wankollo's raised fields fluctuated throughout the growing season. This would certainly have had a very negative impact on the frost resistance functions of the fields, since adequate water levels are the key to maintaining the solar microclimatic effect.

However, I find it hard to believe that the NGO was able to monitor water levels on all four separate plots of raised fields in Wankollo throughout the growing season of 1991. Furthermore, the raised fields built along the river would have had naturally high water levels throughout the rainy season, and did not necessarily need to be irrigated like other plots of raised fields in the community. During 1996–97, while I was living at the Museum in Tiahuanaco, the abandoned raised fields along the river that had been built by the NGO maintained naturally high water levels nearly all year long, except for the very driest winter months of June through August (see Figure 1—Wankollo Raised Fields).

NUTRIENT REGENERATION AND "ECOLOGICAL SUSTAINABLITY"

The retention and recycling of nutrients in the form of "green manure," basically the muck at the bottom of the canals, led many researchers to argue that raised field agriculture was self-fertilizing and regenerative, thus not needing expensive chemical fertilizers. This principal function of the raised fields as a regenerative form of organic agriculture made the fields arguably superior to the modern dry farming techniques of contemporary residents. Additionally, the argument that once the fields were built they could be replanted annually without the need for fallow makes the overall labor estimates drop considerably, after an initially large investment. Some early researchers had even suggested that raised fields might be capable of double cropping (Kolata 1991:110), which Bandy (1999) has called the

"Hyperproductivity hypothesis" for production on raised fields. However, the constraints due to excessive frost causing a very short growing season demonstrate that double cropping simply could not work.

These claims concerning continuous cropping on raised fields are clearly unsubstantiated. Observation of the wide-scale contemporary practice of raised field cultivation in the Bolivian Lake Titicaca Basin shows that most communities in the Catari and Tiahuanaco Valleys that built raised fields with the assistance of the NGO, discontinued cultivation after approximately 2 to 4 years. Furthermore, long-term data on the continuous cropping of raised fields has yet to be provided by either team of researchers on the Peruvian or Bolivian side of the lake.

Many researchers have noted that the water in the canals that surround raised fields not only captures sediments eroding from the planting surfaces, but acts as a natural laboratory for producing "green manure" from aquatic plants, decomposing plant matter, and other nutrients (Carney et al. 1996; Erickson 1988b; Erickson and Candler 1989). Yet none of these researchers have either completed research or made available long-term data on the productivity and fertility of the fields to support their claim. For example, Erickson and Candler (1989:241) write "our (raised field) plots demonstrated sustainable yields; some plots have been continuously cultivated for 6 years without a decline in production." Yet to support this claim the authors reference Erickson's own Ph.D. dissertation, which does not give evidence or data to support the long-term productivity of individual fields. Likewise, Kolata et al. write:

> The raised fields reconstructed by the project are *continuously cropped without the need for a fallow period.* Green manure and organic matter from adjacent canals are routinely incorporated into the topsoils of the planting platforms throughout the field maintenance cycle. This input of new, readily available macronutrients sustains the fertility of the field indefinitely. Technically, there appear to be no physical or agronomic impediments to continuous cropping (Kolata et al. 1986:209).

Again, Kolata et al. (1996) base their report on agricultural production from raised fields during only a single planting season (the season of 1990–91) and they do not give long-term data to support their claim for continuous cropping.

The representation of the fields as ecologically sustainable—that they did not need chemical fertilizers, were regenerative, and capable of continuous cropping—was perhaps the single most significant factor in the promotion of the fields as a form of sustainable agriculture by development workers. Given the Inter-American Foundation's interest in sustainable, local, and grassroots development projects (Inter American Foundation n.d.), it is easy to see how this organization might be interested in a project such as raised fields which was portrayed as a model of ecological sustainability.

Yet it was not just the development agencies and agricultural extension workers who were interested in raised fields for the their continuous cropping and long-term sustainability. This factor was also a major draw for farmers in the Lake Titicaca region who were experimenting with raised fields in the late 1980s and early 1990s. Since normal dry fields can usually only be planted with potatoes for one season followed by a combination of other crops before requiring a period of fallow, the raised fields offered a distinct advantage over normal fields on the condition that they could be continuously cropped. Were raised fields able to be planted annually with potatoes without the need for expensive fertilizers and without requiring a fallow "rest" period between potato crops, they would have been a very appealing cultivation technique for Lake Titicaca Basin farmers. The first reason is obvious, since potatoes are used as a subsistence food rather than as a cash crop, farmers are less inclined to spend money on chemical fertilizers and pesticides. The second reason is that continuous cropping of potatoes would have mitigated the initially high labor demands of building the fields, since maintaining already built fields is much less time consuming than turning over fresh soil for planting regular dry fields.

Yet across the Lake Titicaca Basin, farmers discontinued cultivating individual fields after 2 to 4 agricultural seasons. Kolata et al. (1996:209–10) attribute this to a "cultural resistance" of continuous cropping. Kolata et al. (1996) argue that because local farmers are accustomed to a long fallow system of agriculture, they were reluctant to continuously cultivate the raised fields. Erickson and Brinkmeier (1991) also report that their raised fields on the Peruvian side of the lake could be continuously cropped in potatoes and that the farmers resisted continuous planting of raised fields after one or two seasons. In Erickson and Brinkmeier's (1991) opinion the farmers were basing their judgments on traditional crop rotations and fallow periods, and they were not following the proposed model for raised field farming.

However, Erickson and Brinkmeier (1991) do not provide production data to back up the claim that raised fields can be continuously cropped. Like Kolata et al (1996), the idea of continuous cropping is still a hypothesis that has not been proven. In a later paragraph Erickson and Brinkmeier write, "In our raised field experiments begun in 1981, we found that the first, second and usually third years of potato production are excellent" (Erickson and Brinkmeier 1991:16). However, the reader is left to guess what "excellent production" actually is and also what production was like after the second and third seasons. Given the fact that raised fields were built on nutrient rich and previously uncultivated soil, it is not particularly surprising that they produced one to two, and perhaps even three, years of excellent production. Yet this does not prove that raised fields can be continuously cropped; it only proves that putting land into cultivation that had previously lain fallow for over 800 years is capable of more than a single year and a single crop of potatoes.

I argue that both Kolata et al. (1996) and Erickson and Brinkmeier (1991) have probably overstated this so-called "cultural resistance" to continuous cropping, as well as the long-term sustainability of continuous cropping. Examples from Wankollo demonstrate my argument, particularly since residents had participated in both the community raised fields and private family level raised field cultivation for over four years. Though Wankollo farmers may have initially been hesitant to plant their raised fields continuously—since in their practical experience all fields need a rest between potato crops—they were willing to experiment since their losses were covered by the extra incentives offered by the NGO. These incentives included potato seed, foodstuffs, and hand tools. In Wankollo, it was only after production levels had dropped severely, or failed to produce any crops at all (due to frost), that farmers discontinued potato cropping and began to plant other crops following the normal rotation of crops and ending with the fields reverting back into fallow.

The fact is the development NGO was able to convince Bolivian communities to replant a second season in potatoes, including Wankollo, which clearly demonstrates that they were able to at least initially overcome this so-called cultural resistance to continuous cropping. However, since according to the Wankollo farmers themselves, production on raised fields fell drastically in the 2nd and 3rd years. This demonstrates that the discontinuance of cultivation was simply a response to the declining fertility and production on the raised fields. Furthermore, even where interest existed in replanting the raised fields, local organizers of the work groups complained that they could not organize enough labor to replant the fields by the third and forth years since yields had dropped markedly in the second year.

My own research does not document and record actual potato production on the raised fields, since all the raised fields had been discontinued throughout the Bolivian Lake Titicaca Basin by the start of my field research in 1996. However, on each of the two separate plots of community raised fields, as well as on the four private family-level raised fields in

Wankollo, the farmers recounted a similar pattern of production for each case. In all cases, farmers recounted "normal" (*regular*), "good"(*bueno*), or "very good" (*muy bueno, !super bueno!*) production in the first year of community level raised field cultivation (1990–91). According to farmers, the fields produced so well that this was stated as the primary reason for expanding the community raised fields and for building the three private family-level fields in the following agricultural year (1991–92). Yet in each case, production dropped dramatically in the second year, producing "very little" (*muy poco*), not very good (*no tan bueno, no era muy bien*), or "average" (*regular*) crops of potatoes. However, by the third year all of the raised fields produced very poor harvests and were besieged by frost problems similar to harvests on nearby regular hillside fields. Only the original community raised field plot that was located in a fluvial floodplain in a protected depression of the river produced enough potatoes in the third year so that a handful of residents were prompted to plant a forth year of potatoes. But by the forth year, this field had also completely stopped producing potatoes and was reverted back into fallow.

RAISED FIELDS AS "APPROPRIATE TECHNOLOGY"

In numerous published and unpublished reports, articles, and interviews with archaeologists and development workers, it is repeatedly emphasized that raised fields are built and cultivated strictly with the use of hand tools and manual labor. In proposing raised fields for contemporary development, this theme is emphasized as a selling point under the framework of sustainable development, particularly as a form of sustainable agriculture, since the raised fields do not require costly machinery, chemical additives, or require the use non-renewable resources. Certain models of sustainable agriculture have latched onto the idea that agriculture practices should be developed using "appropriate technologies" that are available to the local farmers.

For example, in the following passage written by the North American archaeologist Clark Erickson:

> The indigenous Andean agricultural tool inventory appears limited in technological complexity, but is more than adequate for the needs of the Andean farmer. Traditional tools include the Andean footplow, hoe, and clod breaker which are still the basic tools today, although stone and wooden blades have been replaced by metal blades (Erickson 1988b:10)

This statement revels assumptions both about the so-called "needs" of the small farmers of the Peruvian Lake Titicaca Basin, but also shows an assumption about the value of labor to smallholder farmers. Erickson writes that altiplano farmers only need "traditional tools" for farming, based on the assumption that manual labor is both widely available and abundant. The implication is that small farmers do not *need* laborsaving technologies and machinery, such as is used in Western style agriculture. However, as I will argue in the following chapter, his assumption that labor is abundantly available for subsistence agriculture is an incorrect assumption.

The Bolivian based promoters of raised field rehabilitation also made similar assumptions about farmers' needs and resources, and what the researchers believe is "appropriate technology" in the Bolivian Lake Titicaca Basin. For example, Kolata et al. write:

> Because the ancient technology of raised-field agriculture is labor intensive rather than capital intensive, its rehabilitation promises to address two critical elements of underdevelopment in the area: inadequate agricultural production and inadequate opportunities for productive rural employment. . . . Our research has demonstrated that . . . the rehabilitation of raised-field plots by a community-based work force is feasible and cost-effective from both economic and social perspectives (rehabilitating raised fields entails only access to hand tools, sufficient improved seed stock, and community cooperation in integrating fields with sources of freshwater such as springs) . . . (Kolata et al. 1996:206) (parentheses in the original).

In the above quote, it seems that what is most glaringly missing in Kolata et al.'s neatly constructed economic model is the *cost of labor*. Like Erickson (1988b), the researchers make the same assumption that labor is in abundant supply on the altiplano. Moreover, another underlying assumption is that increasing potato production would actually improve income and would be "productive rural employment." Yet potatoes in Wankollo are merely a subsistence food, and it is not at all a given that producing more potatoes for a market that must compete with crops from Cochabamba and imports from Chile would actually be "productive rural employment" at all.

For example, from 1987–89, when raised fields were first being implemented in communities on the altiplano, Bolivia imported an average of 376,217 gross kilos of potato per year (Minesterio de Asuntos Campesinos y Agropecurios 1991). As I will demonstrate in chapter 6, agricultural labor, particular adult male labor, is in high demand in Wankollo. Also, since the terms of trade for potatoes are so poor in Bolivia, producing potatoes as a cash crop on the altiplano by increasing household investments in labor is not necessarily a sound economic choice, especially when such labor could generate more income through wage labor else-where.

On the same page as the above quote (Kolata et al. 1996:206) is an illustration of indigenous hand tools, all of which look particularly exotic and prehistoric. One wonders why the authors chose to use a picture of these specific hand tools rather than pictures of farmers using the Western style picks and shovels, such as the ones that the NGO actually distributed to farmers. Both Erickson (1988b) and Kolata et al. (1996) equate appro-priate technology in the case of raised fields with prehistoric indigenous tools, almost as the antithesis to modern Western tools and machinery. They uncritically promote the use of these tools with a simplistic notion that indigenous and ancient tools must equal good agriculture; perhaps since pre-Hispanic Tiwanaku polity was obviously at one time a very pro-ductive agricultural society.

The promotion of hand tools and manual labor for raised field cul-tivation is also argued as economically appropriate and sustainable, since it does not "cost" farm families anything (except labor). Yet in practice, most farmers hired tractors to plow their fields, since one tractor could do in an hour what a farmer with a pair of oxen would take many days to do. Of the 171 individual potato fields for which I have planting data, 125 of them were plowed by tractor (73%). The other 46 fields (27%) were plowed by oxen. It would take a farmer even longer still to do this same work were he using only hand tools like the venerated Andean footplow that promoters of raised fields return to again and again. In Wankollo, most farmers paid to have their fields plowed and thought that the $10 to $14 dollars an hour they paid to have a tractor plow their fields was well worth it, in return for the several days of labor saved. If families have no male residents, they would pay 30 bs. (about U.S. $6) per day to have their fields plowed by oxen. Since plowing a field with oxen that has lain in fal-low for several years can take several days (up to 2 weeks), it was often cheaper and is most certainly easier to have the fields plowed by tractor. No families in Wankollo used footplows, nor did I ever see a footplow in Wankollo.

Of course, in the United States sustainable agriculture using appro-priate technology does not mean giving up the use of tractors and other farm machinery in cultivation. Yet in less industrialized nations many researchers and development workers have proposed that for agricultural practices to be sustainable they must use appropriate technology, this is translated into the use of manual labor and hand tools. It is supposed to make agriculture more readily available to all farmers. In the case of raised fields, researchers and the development NGO workers repeatedly emphasized that the fields be built with manual labor, often emphasizing the use of actual pre-Hispanic tools rather than the Western style tools.

However, as I have argued previously, the theme of appropriate technology in development is almost a self-fulfilling prophecy. It argues that because farmers are already extremely poor and/or only have small farms, they should therefore use only hand tools and manual labor for their agricultural fields. The development logic is that these small farmers either simply cannot afford machinery such as tractors (and therefore they should not use them), or that because their farms are so small, expensive tractors are impractical. An underlying assumption is that poor smallholder farmers have plenty of time to commit to manual labor, when in fact the opposite is probably the case. Examples from production models on raised fields all downplay the investment of labor as insignificant. Yet because the raised fields require strictly manual labor using only hand tools, farming raised fields takes smallholder farmers many more hours of labor to build and cultivate than their regular fields. As I will illustrate in my next chapter, the extra labor required for building and cultivating the raised fields was not well spent once development incentives ceased, and particularly since the raised fields were not continuously cultivated.

Yet by emphasizing sustainable agriculture that uses "appropriate technology," researchers fail to recognize the structural inequalities that make agriculture such a poor venture for generating income in the first place. Were farmers on the Bolivian altiplano offered better prices for their goods, perhaps living standards would improve and tractors and farm machinery would be considered "appropriate technology?" Instead, altiplano farmers suffer from poor terms of trade, and rural development plans that target lowland products. For example, in an interview with a rural development project manager who had been working with USAID in Bolivia since 1989, he indicated that there was very limited funding for agricultural development projects. As he said to me, what limited funds they did have for rural agricultural development at USAID in Bolivia was being directed towards their coca eradication goal of "Coca Free by 2003" in order to "replace coca entirely," presumably with other legal lowland crops.

In most models of production on raised fields, there is an underlying assumption that agricultural labor is abundant on the altiplano, when in fact it is not. I would argue the opposite, that most contemporary farmers are more interested in reducing labor in agriculture, since this frees more household members to seek wage labor in the towns and cities. As previously stated, most farmers in Wankollo paid to have their fields plowed when they were tilling fresh soil that had lain fallow for several

years. Perhaps a more reasonable alternative is to make labor saving technology available on a rental or time-share allotment, rather than simply
propose that these farmers should not have access to such machinery. In
fact, this is exactly how some farmers in Wankollo had their fields plowed,
by paying an agricultural co-op that owned a tractor to do it for them. One
farmer in Wankollo even owned his own tractor and hired out on an hourly
or per field basis to his neighbors.

ACCESS TO LAND FOR RAISED FIELDS

Issues of land tenure and land scarcity have taken center stage in most
ethnographies about rural economy and agriculture in the Andean highlands (Brush 1977a; Carter 1964; Lagos 1994; Sanabria 1993). Therefore,
it is no surprise that archaeologists working in the Lake Titicaca Basin who
supported the raised field project, raised the issue of land tenure and access
to land as a possible stumbling block for the project when put into contemporary practice. Yet access to land in the raised field rehabilitation
project in Bolivia did not pose much of a problem for building the new
raised fields once the communities decided collectively to take part in the
raised field project. I argue that recent out-migration by altiplano residents
has decreased the pressure on land on the altiplano in the Lake Titicaca
Basin, which is one reason why access to land for building raised fields did
not pose a significant problem. The second reason that access to land was
not a significant problem is because the land used for raised fields had previously been considered unsuitable for agriculture. Therefore much of these
types of low-lying marshlands were available to build raised fields. For
example, as I will show in Wankollo, the land deemed most suitable for
raised fields was on communal property formerly only used to pasture animals since it was inundated during much of the rainy season. However,
most farmers have adequate pastures in Wankollo, and taking some communal land to build experimental fields was not particularly problematic.

The first argument I make is that there has been a general decline in
rural population since the agrarian reform, thus reducing the pressure on
land and land scarcity on the altiplano. Though the overall validity of census data in Bolivia is somewhat suspect, I use data from a report compiled
by Crespo (1993) based on the 1950, 1976, and 1992 censuses in this section to illustrate a general trend. I will support the trend demonstrated in
the national census with community level data from Wankollo that verifies
this population movement at the local level.

Since agrarian reform, Bolivia has undergone a transformation from
being a rural country to a population that is predominantly characterized
as urban. According to the census of 1950, the rural population of Bolivia

equaled nearly 2 million persons, while urban centers accounted for only a little over 700,000 (see Table 1—Urban-Rural Population Distribution, Bolivia).

Table 1: Urban - Rural Population Distribution, Bolivia

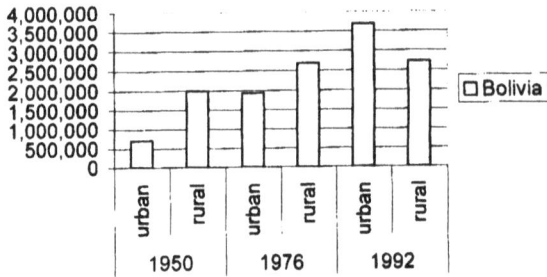

By 1976 the urban population had exploded to nearly 2 million as migrants came from the countryside and into the urban centers, while rural population growth slowed to only 2,690,000. By 1992 rural population had stabilized at 2,730,000 persons, while the urban population continued to expand to nearly 3,700,000 people in the nation's cities. During the period from 1950 to 1992 Bolivia underwent the demographic transition from a predominantly rural society to a newly urban one. From 1950 to 1992, urban population growth registered an incredible 399%, while rural population growth saw a more modest 71% increase (Crespo 1993:23).

Even more telling is the statistical data from the Department of La Paz. This department includes the capital city of La Paz, the eastern yungus valleys, and the northern altiplano, which includes the Lake Titicaca Basin (see Table 2—Urban-Rural Population Distribution, Department of La Paz).

Table 2: Urban - Rural Population Distribution, Department of La Paz

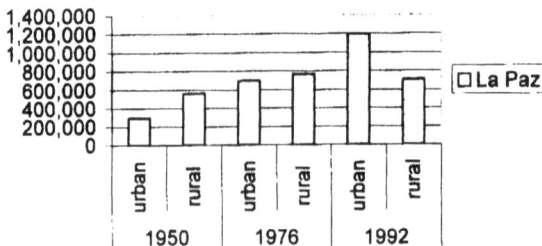

In the Department of La Paz the rural population actually decreased from the 1976 to 1992 as population dropped from just under 780,000 persons down to 710,000 by 1992. Evidence from Wankollo confirms this trend towards depopulation and suggests that this population decrease was the result of large-scale out-migration to the city of La Paz and to other areas of the country.

Other data from Bolivia's national census confirms this trend of increasing out-migration from the rural altiplano. In fact, in the rural altiplano provinces of Bolivia, population has probably been in the decline for 20–30 years. For example, in the rural provinces of Los Andes and Ingavi, which includes the Catari and Tiahaunaco Valleys, population declined –9.84% and –3.85% respectively during the period of 1971 to 1976 (Herrera 1980:29). Based on the 1976 census data, Bartlema (1981) characterizes Los Andes province as a "strong expeller" of population. Bartlema categorizes Ingavi province as a "moderate expeller" of population (1981:15). In either classification, populations are already being recognized as rural populations in decline by 1976. In the census of 1992, population on the rural altiplano continued its downward descent, with an average annual decrease of –0.48% of the rural altiplano population from 1987–1992 (Ministerio de Hacienda 1997:28). Recent projections indicate that this overall decline in the rural population in the province of La Paz will continue through the first decade of the new millennium, from about 733,000 down to about 680,000 by 2010 (INE 1997:61).

To get a general picture of the population of Wankollo from the agrarian reform to 1996, I have two primary pieces of evidence: the agrarian reform map that lists the number of households, and census data from a household census I conducted in October and November 1996. Secondary evidence includes information and oral histories from community members about the population and history of Wankollo.

The agrarian reform map (see Figure 4—Agrarian Reform Map) lists the number of households in Wankollo according to their land grants. The map is dated July 22, 1960 and it lists a total of 121 households, though of these 37 were listed as "newly gifted" (*nuevo dotados*) households that only received between 3 to 5 hectares of land. If we disregard these newly titled households (mostly these new titles went to adult sons of residents), that still left a total of 84 households whose land grants ranged from 3 to 40 hectares. Of these 84 original households, the land grants were a bit more generous in comparison to other northern altiplano communities (Carter 1964) and all but 14 households received at least 10 hectares of land.

By the time of my household census in 1996 the total number of households in Wankollo was 90, thus the number of households had only increased by 6 since the agrarian reform. While conducting the census, my assistant and I walked over the entire range of the community and accounted for the residents of every structure that was inhabited in the community. In addition to the 90 full-time households, there were 6 households whose "owners" lived permanently in the city of La Paz, though they returned periodically to maintain a small number of agricultural fields. Of these, 2 households had owners who lived in La Paz and were occupied by a guardian (*cuidador*), who did not participate in the community *sindicato* (local community union) or other community activities, and was not considered a member of the community. Given that the rural altiplano of La Paz has seen a net expulsion of population, these data from Wankollo fit the overall pattern for the region.

Because population has remained relatively stable in the community, and with most households receiving over 10 hectares of land during the agrarian reform, land scarcity in Wankollo is a less critical issue. Of the land that was distributed with the household *sayañas,* nearly all were located on the *pampas* (the flat basin floor of the altiplano). Very few families (6) were granted *sayañas* that included a portion of hillside land. The average amount of land available per household, including the community *aynoka,* was over 20 hectares. Of course this does not include rights to land by family members who live away from the community, though it does demonstrate that access to land was less critical. In nearly all of my interviews with farmers, they did not regard access to land for agriculture to be a significant problem. Only in a few rare instances did a farmer tell me that he was unable to plant more crops because he did not have suitable land or enough land. As I will demonstrate in the following chapter, farmers throughout the community told me that the primary factor limiting the number of fields they cultivated was the amount of labor they had access to, and not the amount of land.

However, farmers in Wankollo were still very interested in a technology such as the raised fields that might give their *pampas* fields an extra advantage in agricultural production. Because most farmers did not have any hillside fields to cultivate, the community had an added generalized risk of frost because they had little diversity of field types. Thus, adding raised fields to the field type regime was initially appealing to members who wanted to build the fields to artificially create a third microclimate with which they could diversify their production strategy. The production strategy of diversification across different microclimates is common in Andean highland communities (Carter 1964; Lagos1994; Sanabria 1993), though on the northern altiplano of Bolivia the agrarian reform terminated any remaining access to lower altitude fields (Birbuet 1992).

In the previous paragraphs, I demonstrated that due to the large size of the original land grants, which when combined with high out-migration from the community, resulted in access to land being a less critical issue in determining how many fields one planted and one's ability to access land for building raised fields. The second factor in accessing land for raised fields was that the land considered the best for building raised fields was land that had previously been considered unsuitable for agriculture and therefore only fit for pastures.

The ancient remains of raised fields are found throughout the Lake Titicaca Basin on the low laying areas of the altiplano floor (Smith et al. 1968). These areas are typically marshy lands, often times seasonally inundated by water during the growing season. Generally they have been considered only fit for pasturing animals and not fit for agriculture, both because of the marshy conditions and also due to the exposure risk to frost on the low spots of the altiplano floor. Access to these lands and tenure vary both between Bolivian and Peru, and between different communities. In some communities individuals "own" these lands as a part of their household *sayaña* and have sole access to it, while in other communities it is considered communal pastures (*aynoka*) with a variety of tenure rules pertaining to it.

Access to land for raised fields has different variables for both the Peruvian and Bolivian sides of Lake Titicaca, due to the different agrarian reform measures in each state. Erickson and Brinkmeier (1991) report that recent land reforms in Peru returned much of the land on which the ancient raised fields are located back to the indigenous communities with the stipulation that these lands must be farmed communally. According to the authors, the communities are eager to put the land under cultivation since this clearly demonstrates land ownership by the community. Raised fields offer a way to cultivate these often wet and inundated lands that had previously been used only for pastures. Cultivation of land, particularly the highly visible raised fields, is a political symbol of ownership and is an important way that communities can demonstrate their collective rights to land (Erickson and Brinkmeier 1991:9). However, the authors do note that raised fields are not compatible with the use of these fields as pastures since grazing animals, particularly cattle, would erode the raised platform beds.

In the case of Wankollo, and in most ex-hacienda communities such as the majority of communities in the Catari Valley, there is little communal land available for community-wide cultivation. The communal land in Wankollo is called the *aynoka*. The aynoka is communal land because it is "owned" by the community. Carter (1964) notes that these communal *aynoka* lands where probably cultivated using a sectoral fallow system, where a whole *aynoka* is cultivated in the same rotation (usually one year of potatoes, followed by quinoa, followed by barley, followed by a fallow period of indeterminate length).

However, in Wankollo the *aynoka* is no longer (and perhaps it never was) cultivated using a sectoral fallow system. Instead, the community *aynoka* of 145 hectares was partitioned into ¼ hectare plots with most families having access to between ½ to 1 hectare of these lands. Today in Wankollo, access to specific *aynoka* plots is inherited much like private property. However, unlike private property, if this land is uncultivated (excluding normal fallow periods) it can revert back to the community and be redistributed. The "owner" of any particular *aynoka* plot can choose when and what crop to plant, as well as make arrangements for share-cropping or renting the land, though he or she cannot outright sell the *aynoka* plot.

Based on the agrarian reform map for the community of Wankollo, there are 98 hectares of land that was termed "uncultivable and rivers." Much of this land is the low laying area in the streambed that runs through the middle of the community. This land has communal access by all community members and is not cultivated. Community members primarily pasture animals in the area. Like other communities, this "uncultivable land" was the best location to build raised fields in Wankollo. The community level raised fields were built in a particularly flat and wide stretch of the streambed, with one section of raised fields close to the church and community buildings and another near the schoolhouse. Since most farmers had enough land to pasture their animals, taking a few hectares of land that was unsuitable for agriculture did not pose a problem for residents of Wankollo. In fact, in the second year of cultivation three individual families built raised field on their own private property and they built it on land that had been used for agriculture in the past. These farmers saw little problem with taking a ¼ to ½ hectare of their own land to build the experimental raised fields.

As I have argued in this section, access to land was not a limiting factor for residents of Wankollo to practice raised field cultivation. Given generalized population declines across the rural altiplano, it very likely was not as important an issue as one might suspect it to be since it has become less of a scarce resource with declining populations. Though, given the indigenous peasantry's historic and tenacious battle to preserve their land base, negotiating access to land for raised fields still posed some initial impediments to constructing raised fields.

CONCLUSION

In practice, the Bolivian raised field rehabilitation project did not live up to the expectations of the development group and the archaeologists based on the models of production devised by archaeologists and other researchers. In terms of the frost mitigating benefit of the raised fields, researchers have determined that the water in the canals do retain some degree of heat, which does offer some protection to crops on raised fields. However, it is unclear how much extra protection this offers in the long-term, since current models do not factor in how much frost protection benefits are derived from simply putting nutrient rich virgin soils into cultivation. It is likely that any field that has nutrient rich virgin soil is going to offer some extra protection to crops, simply because healthy plants thrive better and have a stronger resistance to frost.

Second, in practice the raised fields have not proven to be regenerative or self-fertilizing, and therefore they have not been continuously cropped in potatoes. Across the Tiahuanaco and Catari Valleys, raised fields built by the NGO were abandoned after 2 to 4 years. A majority of the fields were built by the NGO between 1989 and 1992, and by the time of my own arrival in 1994 most of these fields had already been abandoned. Though archaeologists working in both Peru (Erickson and Brinkmeier 1991) and in Bolivia (Kolata et al. 1996) have argued that there is a cultural resistance to continuous cropping, at least in Wankollo the NGO was able to overcome this resistance to continuous cropping. As I have argued, it was only after production had declined considerably, or the raised fields had stopped producing any crops at all, that the fields were returned to fallow or were planted in barley.

The third theme considered the nebulous category of raised fields as "appropriate technology." Raised fields as appropriate technology has been represented through the promotion of the fields as indigenous, needing only hand tools and manual labor for construction and cultivation. It was been argued by archaeologists that raised fields are appropriate technology simply because they are not capital intensive and do not use expensive machinery, fertilizers, or pesticides. I question the very use of the concept of appropriate technology, since what is considered appropriate for a farmer is based on structural inequalities imposed by a long history of oppression and exploitation. Hence, by promoting a development project considered appropriate technology for a small farmer with little or no cash income derived from farming simply serves to maintain current inequalities. Representations showing "indigenous" farmers using primitive hand tools and working together in workgroups are not necessarily depicting indigenous Andean "traditional agriculture." Such images of traditional agriculture is a result of a history of oppression and exploitation, hence a farming system that continues to promote these Andean traditions is in effect promoting the continuation of structural inequalities.

And lastly, I counter the concerns that access to land poses a significant problem for constructing raised fields. I argue that out-migration has reduced pressure on access to land, particularly for the low-laying land that are the optimal location for building raised fields. In Wankollo, farmers repeated these same sentiments and usually reported that they were unable to intensify their agricultural production due to a lack of labor rather than insufficient access to land.

In the following chapter I address labor in agriculture and in the raised field rehabilitation project. The raised field rehabilitation project, which was based on archaeological models of raised field production, was fabricated on the assumption that labor is abundant and in unlimited supply on the altiplano. None of the models of production on raised fields places much value on agricultural labor. All assume that once the superiority of raised fields is demonstrated, access to labor for building and cultivating raised fields will not be an issue. Few of the models even bother to discuss access to appropriate labor.

Given that raised fields were not continuously cropped (regardless of any cultural resistance that researchers have argued), I argue that once extra incentives were discontinued, the cost of labor for building raised fields was simply too high. I demonstrate that raised field agriculture is labor intensive, particularly relative to current agricultural practices on the altiplano. Therefore, since production on the fields was not continuous, the initial large investment of labor and the leadership required to organize this labor was not forthcoming following the withdrawal of NGO leadership and financial incentives.

CHAPTER SIX

The "Myth of the Idle Peasant" Revisited: Access to Labor for Agriculture

S OME YEARS AGO STEPHEN BRUSH WROTE AN ARTICLE TITLED "THE Myth of the Idle Peasant" (1977b). In this article, he argued that in the highland Peruvian community in which he conducted field research, there was full employment of the community population. Further, he argued that there was no unemployment or underemployment in any segment of the community. His argument was counter to economic and development theory at that time, which generally maintained that rural overpopulation had caused generalized underemployment in the Peruvian countryside. Since the publication of his article, other anthropologists have added to the debate over rural underemployment, putting to rest the myth that there is generalized rural underemployment (Netting 1993).

In his argument, Brush (1977b) makes the distinction between strictly agricultural tasks and other types of work that farmers do on the farm. However, he does not evaluate the role of off-farm labor. In the case of Wankollo, which sits adjacent to the town of Tiahuanaco and is only two hours away by bus to the capital city of La Paz, off-farm labor is a very important factor in evaluating access to labor for agriculture. Any rural development project that increases the demand on agricultural labor must calculate in this cost since labor is often diverted from agriculture into off-farm activities and wage labor. As I have already demonstrated, archaeological and development models of raised field production have consistently undervalued the cost of labor. Previous models of raised field production have assumed that access to labor posed little problem for farmers, and they have not adequately assessed the cost of accessing that labor.

In the previous chapter, I outlined several production factors that influenced the building, continued cultivation, and production on raised fields in the Bolivian Lake Titicaca Basin. This chapter examines access to labor for agriculture, for both raised fields and regular dry fields, which I argue is the primary factor limiting the contemporary usage of raised fields in agriculture in Bolivia.[1] I argue that labor is not an abundant resource on the altiplano, as so often assumed in models of raised field production and as I demonstrated in the previous chapter.

Access to labor for agriculture is important because of three inter-related issues. First, as I have demonstrated in the previous chapter there is a general trend of out-migration from the rural altiplano areas, which has resulted in a decrease in population in the rural areas of the province of La Paz. This generalized decrease in rural population also suggests a decrease in labor available for agricultural tasks.

Second, as I will demonstrate in this chapter, data from Wankollo shows that women outnumber men roughly 4 to 3, increasing the demand for male labor. In Wankollo, adult male labor is usually "paid" labor (either in cash or in kind), particularly during the harvest. In addition, many men work off-farm for much of the year, increasing the responsibili-ties of women in agriculture. I use census data from Wankollo to demon-strate the distribution of sex and age across the community, and specific household cases as examples of this trend.

Third, the construction and cultivation of raised fields was signifi-cantly more labor-intensive than the regular dry fields, adding a further demand on household labor. The pay-off for agriculturalists building raised fields was supposed to be the substantial increases in production that would make this labor-intensive agriculture worthwhile to smallholder agriculturalists. Given that labor is in high demand for agricultural tasks and in limited supply, I argue that the amount of labor required to build and maintain the raised fields is relatively higher when compared to regu-lar dry field farming on the hillsides and flatlands. If raised fields cannot be continuously cultivated, as was illustrated by the abandonment of the sys-tem after 2 to 4 years of cultivation in communities that had built the fields with the help of the NGO, then the initial high labor costs of building the fields are not balanced out with added years of cultivation.

Finally, I argue that in the contemporary social setting within lake basin communities such as Wankollo, the problem of organizing large labor groups for performing a common agricultural task is considered an additional burden. Organizing large groups for agriculture is not a simple task of calling on "traditional labor organization," as some models for pro-duction on raised fields have promoted. I also give evidence that reciprocal labor exchanges were not common in Wankollo in 1996–97, particularly the equal exchange of labor for labor (*ayni*). Instead, families prefer to use household (*familiar*) labor as their first choice of agricultural labor, followed by different forms of sharecropping and "paid" labor, both in cash and in kind. This pattern is also related to the increased demand on male labor. However, informants also report that there has been decreasing production on all agricultural fields in the community in recent decades, thus community members prefer to be paid in kind for their labor, particularly during the harvest.

Similarly, Wankollo residents also find the task of organizing work groups onerous and difficult, especially given declining production returns on raised fields previously built with the aid and incentives of the NGO. Once organizational leadership was withdrawn and incentives were discontinued, the building and cultivating of new raised fields was also discontinued on the Bolivian altiplano.

Labor estimates and labor data for this chapter are based on semi-structured interviews during the sowing and harvesting of the potato crop in Wankollo, as well as informal interviews and participant observation. Potatoes are the principle subsistence crop on the northern altiplano of Bolivia, as many other Andean crops, including maize, cannot be cultivated at such high altitudes. The labor requirements for cultivating and harvesting potatoes outweigh other crops, though a number of other indigenous crops and barley are cultivated. For quantitative data on types of labor and frequency of labor types, I focus on the potato harvest because it requires the most labor within any given segment of the agricultural season. For the harvest data, I draw on a total of 50 semi-structured interviews with households in Wankollo, in which I collected labor data for each field of potatoes that was cultivated by the household. Though my sampling strategy would be considered a convenience sample, I was able to interview well over 50% of the households in Wankollo with households representing both the *"ex-comunario"* and the *"ex-colono"* population. In addition to community members,[2] I interviewed one household that resides within the community but did not participate in the social and political life of Wankollo and was not considered a "community member." This household was from another community and acted as guardians for local absentee owners. I also interviewed one landowner who lived in La Paz and also did not participate in the social and political life of the community.

During both sets of interviews (the planting and the harvest), I consistently probed informants on the types of labor that they utilized to cultivate their potatoes fields. When it became apparent during the planting interviews that reciprocal labor exchanges were not common, I began to ask very specific questions regarding the use of *ayni* labor exchanges (rather than simply relying on open-ended questions). In my interviews, Wankollo residents could all explain to me what *ayni* labor exchanges were and when they were likely to be used. However, they said that this type of labor exchange had been used more in past generations. I will address the likely reasons why the exchange of labor for labor via *ayni* was not common in Wankollo during the 1996–97 agricultural season in more detail later in this chapter.

As stated above, labor data was collected for each individual potato field that was planted by each household that was interviewed. Labor data was gathered for the number of persons and number of days that each person participated in the planting and the harvest. For example, if a husband, wife, and one teenage daughter harvested one potato field over a period of 3 days, this counted for 9 person-days of household labor. If they were assisted for two of those days by the wife's mother-in-law who was paid in kind (*minka*), the total number of person-days for harvesting the field would be 11 person-days, with 9 person-days from household labor, and 2 person-days of *minka*.

ORGANIZING AGRICULTURAL LABOR

The primary and preferred source of agricultural labor for all households in Wankollo, and in other Andean communities, is familial (*familiar*) labor (Brush 1977a; Chibnik and Jong 1989; Guillet 1980; Sanabria 1993). Familial labor is unremunerated household labor that is supplied by the immediate family of the household, which usually includes the husband, wife, and their children. It also may include any other household dependents such as adopted or illegitimate children, as well as grandchildren that live in their grandparents' household. In the potato harvest of the agricultural season of 1996–97, 74.2% of labor deployed was familial labor (see Table 3—Distribution of Harvest Labor in Wankollo).

Table 3: Distribution of Harvest Labor in Wankollo

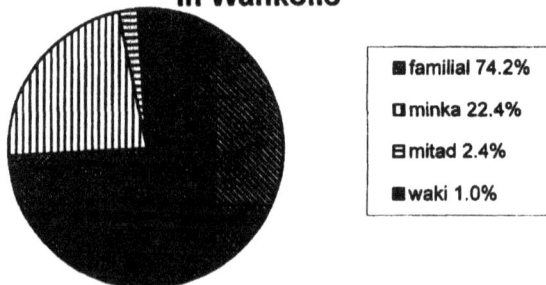

■ familial 74.2%
□ minka 22.4%
⊟ mitad 2.4%
■ waki 1.0%

Reciprocal Labor

As one moves outside of available household labor and to immediate and extended kin, there are different types of reciprocal labor arrangements that are made. The first type is an open-ended form of "generalized reciprocity" (Sahlins 1965). In Wankollo, it is called by the Spanish names of *colaboración* or *ayuda,* or by the Aymara word *yanapa.* This form of generalized reciprocal labor is primarily between close kinsmen, including affinal and fictive kin, though for projects of short duration it may be carried out between neighbors, such as the plowing of potato fields for harvest. The repayment of generalized reciprocity is open-ended, with no specified terms of kind, amount, or time (Brush 1977a), though as Sanabria (1993) notes the obligation to repay this labor in some form is very strong within and between households. In Wankollo during the 1996–97 agricultural season, *ayuda* labor was not utilized for the potato harvest, except for the occasional loan of a pair of oxen to open up rows of potatoes to be harvested. However, *ayuda* labor was prevalent in other agricultural tasks, such as sowing potato fields or for harvesting and threshing grain.

A second form of reciprocal labor is referred to by Sahlins (1965) as "balanced reciprocity." It is known as *ayni* in most Andean communities (Buechler 1969; Carter 1964; Erasmus 1956; Sanabria 1993), including Wankollo. *Ayni* is an even exchange of labor for labor, usually between immediate and extended family members. Records of who owes labor and who is owed labor are scrupulously maintained, either in memory or in writing (Carter 1964; Sanabria 1993). All of the community members I interviewed recognized this type of labor exchange and said that it had been practiced in the past. However, there were no recorded instances of *ayni* labor exchange during the potato harvest in the 1996–97 agricultural season. This is not particularly anomalous, since *ayni* labor is less significant during the harvest than at other times of the agricultural cycle (Lagos 1994; Sanabria 1993).

A third type of reciprocal labor is called "festive labor" (Erasmus 1956) and is often called a *faena* in the Andes (Brush 1977a; Erasmus 1956; Guillet 1980). Festive labor includes the providing of large quantities of food and drink, and sometimes music, by the host of the labor party. A festive atmosphere pervades such gatherings, while workers work for the host of the festive labor party. *Faenas* are not balanced reciprocity, because the host is not obligated to reciprocate the labor given to him, though he or she is expected to participate in the festive work parties of others (Mitchell 1980). There were no *faenas* or other types of festive work parties for agricultural labor in Wankollo during the planting and harvest of 1996–97.

Nonreciprocal Labor

Aside from the use of household labor in Wankollo, which dominated all forms of labor deployed for agriculture, they were several types of arrangements for labor that can be categorized as non-reciprocal. Most non-reciprocal labor exchanges in Wankollo were non-wage exchanges that were paid for in kind, rather than in cash. The following discussion is in order of their prevalence of use in the community. The first type of non-reciprocal labor was the use of *minka*, also known in Wankollo by the Spanish term, *contracto*. *Minka* is paid labor, paid for in cash or in kind. It is a common throughout the Andes, though it is often known by different names. For the rural Aymara speaking communities on the altiplano of La Paz, both Carter (1964) and Birbuet (1992) describe *minka* as contract work that is paid for either in cash or in kind, with the latter usually being paid for assistance given during the harvest.

In other areas or Bolivia, *minka* is communal work (Lagos 1994). Labor that is paid for in kind is known as *paga,* and labor paid for in cash is known as *jornal* (Sanabria 1993). In Peru, Brush (1977a) describes *"minga"* as labor that is paid for in kind, while labor paid for in cash is called *jornal.* However, as I have noted previously, both myself, Carter (1964), and Birbuet (1992) have found that in Aymara-speaking communities on the northern altiplano of La Paz, *minka* describes both types of labor arrangements, either in cash or in kind.

After household labor, *minka* was the most important form of labor arrangement in Wankollo during the potato harvest of the 1996–97 agricultural season. During the potato harvest, all *minka* labor was paid for in kind (potatoes). One day of *minka* labor during the potato harvest was typically paid for by ½ quintal of potatoes. Often *minka* labor was also combined with other arrangements for access to land and seed. For example, a person who does not have enough land might arrange to sharecrop a field with another resident, while also arranging *minka* labor for help during the harvest. Likewise, a resident who does not have enough seed, but does have sufficient land, can arrange to have another resident plant his field in return for 50% of the harvest (this is called *waki,* see below), and he or she may also arrange *minka* labor for the harvest of the field.

During the potato harvest of 1997, 22.4% of labor deployed was *minka* labor. *Minka* labor for the potato harvest in 1996–97 was always paid for in kind, with men and women receiving the same rate. During the altiplano harvest in May of 1997, ½ quintal of potatoes sold in the town of Tiahuanaco for 30 bs. (about U.S. $6), while in La Paz it may have sold for up to 40 bs. (U.S. $8). In comparison, to contract a man to work with a pair of oxen for the day during planting cost 30 bs. per day, though sometimes a row of potatoes at the harvest was also included.[3] Contracting a pair of oxen for barley and other crops was also paid a flat cash rate of 30 bs. per day. There is an association of men in the community who hire out for this work with their own oxen. They organized into this association and set the rate at 30 bs. per day because adult male labor is relatively scarce and it is usually only male labor that is used for plowing fields with oxen. I discuss male labor scarcity further in the following section. Contracting *minka* labor for other agricultural tasks is paid for at varying rates, though usually somewhat less than the price of labor during the potato harvest.

The following four types of labor exchange arrangements were not statistically significant. However, there were instances of their use in the potato harvest of 1997. The most prevalent of these was a labor arrangement known as *partida* or *a partir* (Carter 1964). Similar to sharecropping, *partida* arrangements are made when one household has land but not enough labor. The landowner recruits labor for cultivating the field and occasionally helps to cultivate the field with the laborer or laborers. The landowner and laborers split the harvest between themselves. Of the 50 households interviewed for the 1997 potato harvest, there were 6 instances of *partida* labor exchange.

The next type of non-reciprocal labor arrangement is known as *waki*. Carter (1964) previously described *waki* labor exchanges in Aymara communities of the Lake Titicaca Basin following agrarian reform. *Waki* is an arrangement used when a landowner does not have enough seed to plant his fields and therefore he cooperates with another community member who has seed. In such cases, the landowner prepares the field for planting and provides half of the manure. The seed-holding person comes on the day of the planting with the seed and half of the manure. Together they sow the field. The harvest is divided into every other row and each participant harvests his own rows. Carter (1964:51)) notes that this exchange can also be used by a farmer who has adequate seed but wishes to acquire a specific seed variety from another farmer. In the 1997 potato harvest, there were 5 instances of *waki* labor exchange.

The forth type of non-reciprocal labor exchange was known as *mitad*. *Mitad* labor exchanges were usually between immediate kin, such as siblings. Usually a pair or more siblings work one of their parents' fields together, with equal inputs of labor, seed, and fertilizers. At harvest, they split the crop equally between themselves. In the 1997 potato harvest in Wankollo, there were 4 instances of *mitad* labor exchange.

The fifth type of labor was *sajta*. Carter (1964:50) had a similar category that he labeled *sattakha* and he called this form of labor "the peasant's social security." In Carter's experience, *sattakha* arrangements were made for landless and incapacitated members of the community, or for orphans. In Wankollo, there was one instance where an older widow in the community gave labor to her son and in return he cultivated 2 rows of *habas* in his field for her. Another widow assisted in the potato sowing of her son. In return, she was given a few rows in the field for herself and for which she was responsible for harvesting.

The final type of labor in Wankollo was what Guillet (1980) termed "custodial labor." Different from previous forms of labor, custodial labor "is characterized by an obligation to participate based on differentially ascribed power of local authorities. Sanctions, both formal and informal, compel participation" (Guillet 1980:154). Examples of custodial labor in Wankollo include community work projects such as the fixing of the roads and community buildings. However, these instances are not *faenas*, because households are required to participate and there is no sponsor who provides food, beverages, and coca. Recruitment of custodial labor is by collective decision either made consensually or through an individual who is delegated authority.

Group labor recruitment for community level raised fields in Wankollo fell into the category of custodial labor. However, the obligation to participate in raised fields seemed a bit more lax than the obligation to participate in other group labor projects, such as the fixing of the community road after seasonal flooding had eroded parts of the road. The family level raised fields were organized along the lines of *partida* labor exchanges, with an individual landowner who organized labor from extended family and neighbors. The participants received part of the harvest from the individual family level raised fields in return for their labor.

The preceding section demonstrates that reciprocal labor was not at all common in Wankollo in the 1996–97 agricultural season. Further, farmers who participated in the raised field rehabilitation project repeated to me similar complaints that organizing labor for building and cultivating raised fields had been problematic. No examples of balanced reciprocal labor exchanges *(ayni)* were found during the agricultural season of 1996–97. Farmers did not care to participate in these exchanges, in part due to the obligation they incurred to repay this labor at a later date. Instead, Wankollo farmers clearly preferred to pay either in cash or in kind for the labor that they required to harvest their potato fields, if that labor was not available from the household.

When I asked my assistant, who was himself a moderately prosperous farmer in Wankollo, why he thought that people did not use reciprocal labor very often, specifically *ayni*, his opinion was twofold. First, he said that there were fewer men in the community (*por falta de hombres*), and "when there are more men, there is more collaboration. Because there are not many men in the community, one needs to pay them [the men] to work." His second reason was due to productivity (*por falta de producción*). He said that, "when there was sufficient production [in the community], the people don't need to work for *minka* [potatoes]." Both of these reasons are interrelated and both are linked to my next argument, that there is markedly less male labor available for agricultural work in the community.

LABOR AVAILABILITY IN WANKOLLO

In the following section, I demonstrate that labor is a limited resource and that Wankollo families experience labor shortages throughout the agricultural season. In particular, I focus on adult male labor in Wankollo, which is at a high premium and usually must be paid for in cash or in kind. Men in Wankollo did not work for *ayni* labor exchange, though they did assist with *ayuda* labor to close relatives or on projects of short duration.

However, it should be noted that male and female agricultural labor is equally valued in the Andes (Hamilton 1998), and my focus on the lack of male labor is merely an indication of a general labor scarcity. This is not meant to imply that women are not equally as capable of managing households and providing agricultural labor. In fact, this is exactly what they did do since they provided the majority of the labor for agriculture in Wankollo. Furthermore, in regards to raised fields, it has already been demonstrated that raised fields can be built and managed by women, most notable with the *Centro de Madres* group from the community of Lakaya (Kehoe 1996).

If we look at the age and sex distribution of the 401 residents of Wankollo, there are two indicators of the amount and quality of labor available. First is the age distribution of Wankollo residents (see Table 4— Age Distribution in Wankollo). In Wankollo households, children under 18 years of age accounted for 199 of 401 residents, or just under 50% of the resident population. Within the adult population, 34 residents are 65 or older, about 8% of the population. That leaves a total 168 adults between the ages of 18 and 64 to do the majority of the agricultural work.

The sex distribution in Wankollo is also interesting, because females significantly outnumber males across all three age categories (see Table 5— Sex Distribution by Age in Wankollo). However, most notable is the difference between the number of males and females within the adult age category of 18 to 64 years. In this age category, females outnumber males 96 to 72. This equals about 4 adult women for every 3 adult men. As I will

Table 4: Age Distribution in Wankollo

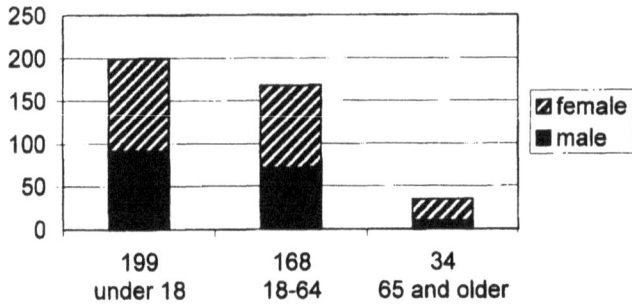

| | 199 under 18 | 168 18-64 | 34 65 and older |

explain below, the ratio of females to males in the community on any given day is typically even higher. This is because my census counted as part of its resident population, those men who actually spent much of the year residing away from the community.

Table 5: Sex Distribution by Age in Wankollo

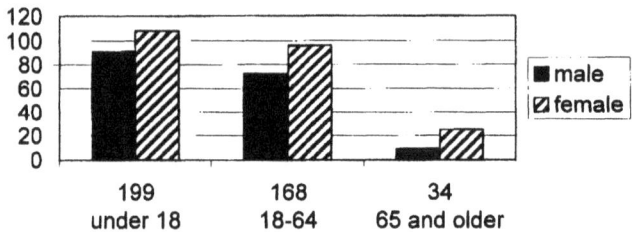

| | 199 under 18 | 168 18-64 | 34 65 and older |

Going through the census data from Wankollo household by household reveals that 25 of the households out of a total of 90 households in Wankollo (27.7%) do not have *any* adult men residing in the household, not even on weekends or for part of the year. These households had no men who could be called upon for general agricultural labor and they were required to pay others in the community to do some of the heavier tasks relegated to men, such as plowing. Further, of the 72% of households with adult men, many of these men worked off-farm during the year. Some men worked in the cities of La Paz or El Alto all year long, Monday through Friday, often living in the city during the week and returning to the coun-tryside on weekends to participate in community activities. This was the case for the community leader of Wankollo (*secretario general*), who worked during the week as a tailor (*sastre*) in the city. Another man worked

6 days a week at the site of Tiwanaku. A number of men also sought occasional work in the town of Tiahuanaco. There were several men in Wankollo who were members of the local union of archaeological workers, called *maestros,* who worked off and on for various archaeological projects. Three *maestros* from Wankollo had worked on projects in the 1996 season alone. However, these projects tended to interfere less with agriculture than other jobs or types of off-farm labor, because excavations usually take place during the southern hemisphere winter months of June through August (after the potato harvest and before the new spring planting).

For example, Juan, a Wankollo resident aged 44, worked at the site of Tiwanaku 6 days a week, leaving only his wife and children under the age of 14 to take care of the majority of the farm work. Four of Juan's older children lived in La Paz, though some of his children returned to Wankollo to assist during the harvest and worked for *minka.* These children received potatoes in return for their labor during the harvest of their parents' fields. Juan and his wife had approximately 1½ hectares of potato fields spread out between three separate plots. The children who lived in La Paz were not considered familial labor because they no longer resided in the household and were paid for their labor with potatoes under *minka* labor exchanges. Though they may have felt obliged to help their parents during harvest, nonetheless they still received the going rate of ½ qq of potatoes for their efforts.

Another 3 men were working full-time in La Paz and returned to the community only on the weekends, with many more men from Wankollo who worked in La Paz on an occasional or opportunistic basis. For example, Victor, aged 38 and the leader of the community *(secretario general),* worked 5 days a week as a tailor making suits in the city of El Alto. He returned to the community only for the weekends, while his wife managed the farm and recruited labor for help during peak labor shortages. Victor had six children, though the three oldest lived in La Paz and rarely came to visit the farm. The youngest three, ages 8, 10, and 12, lived on the farm, went to school in town, and helped with the agricultural tasks when they were not in school. His wife, Rosa, did the majority of the agricultural work on their 1¼ hectares of potato fields. For the potato harvest of 1996–97, Rosa worked on two of her ¼ hectare fields with her sister and mother-in-law, who both worked for *minka.* She split the labor on one ¼ hectare field with her sister who worked with her by *mitad.* Finally, Rosa harvested the family's last ½ hectare field located near the family house compound alone by her self.

At least another three men in Wankollo worked in various other rural locations, including the eastern sub-tropical region. For example, David was aged 32 and worked 20 days per month in the eastern sub-trop-ics. He worked for an NGO as a naturalist and returned to Wankollo when he could on weekends and holidays. He had five children all under the age of 12. He and his wife had only 2 small fields of potatoes in 1996–97. The first ¼ hectare field they planted through *waki* exchange because they did not have enough potato seed; this meant that their own share of the har-vest was reduced to only half of the field. They harvested their half with the assistance of 5 persons from their extended family to whom they paid in kind via *minka*. The second field that the couple cultivated was ¼ hectare, which they harvested together with the help of 5 extended family members that they paid in kind via *minka*.

The previous examples illustrate the various jobs and types of work that employ men in Wankollo away from their farms and the community. Of course, many more men spend the majority of their time in Wankollo though they do participate in wage labor on an opportunistic basis, such as working for archaeologists (or the occasional cultural anthropologist) or working in the town of Tiahuanaco. There are fewer opportunities for women to do wage work in the countryside, though one young woman did work selling locally made tourist goods at the archaeological site of Tiwanaku. This section has served to illustrate the various types of off-farm wage labor that draws labor away from agricultural work in the commu-nity of Wankollo. The following section examines the amount of labor required for raised field cultivation and compares this with regular dry farming techniques.

LABOR REQUIREMENTS FOR RAISED FIELDS AND REGULAR POTATO FIELDS

In this chapter, I make the argument that the cost of labor for raised fields was simply too high relative to the productivity of the fields, thus raised fields were not a sound choice for households that depended on off-farm income. To make this argument, I will demonstrate in the following section that raised fields were more labor intensive than regular hillside or flatland fields. Further, though production per unit of land was higher on raised fields, the production per unit of labor was not. In terms of labor, raised fields do not produce enough over the long-term to outweigh the consider-able investments in labor that are required to build the fields.

It has previously been argued by Erickson (1992a, 1993) that raised fields are not labor intensive relative to regular fields, since after an initially high labor investment in building the raised fields they required little maintenance and could easily be replanted each year. Erickson writes:

> Raised fields are not necessarily labor intensive. . . . The initial construction of raised fields—digging canals and transferring soil to construct the platforms—involves a relatively large labor input (in comparison to traditional *wacho* lazy bed construction). . . . Labor input on raised fields becomes almost negligible when labor and production are considered over the long term. Raised field agriculture is efficient and sustainable because fields can be farmed continuously with high productivity for many years (Erickson 1993:405, parentheses in the original).

Erickson's argument is that in the long-term raised fields are not more labor intensive because they can be continuously cropped. Though farmers put in an initially high investment, Erickson (1992a, 1993) maintains that the cost of maintenance and cultivation on raised fields each year is minimal in contrast to the laborious task of cultivating new fields year after year that had been in fallow.

However, given that raised fields have not been proven to be capable of continuous cropping, the cost of labor for raised fields increases considerably. My own data collection methods for production on raised fields in Wankollo relied primarily on oral histories and recollections about production on raised fields. My goal was to reconstruct production and labor deployment on the raised fields in Wankollo. Admittedly, this is not the most effective way to measure labor and agricultural production since recollections of seed, fertilizer, and labor investments would probably be distorted in hindsight. Therefore, I also rely on Erickson's (1985, 1988a) labor data for the reconstruction and production on experimental fields in Peru. I supplement this with production data on regular dry fields and from recollections about production on raised fields in Wankollo. It should be noted that different researchers give varying data on the amount of labor required for construction and production on raised fields. It is very likely that labor requirements did vary considerably between communities, due to factors such as the compactness of soil, moisture level of soil, type and quality of labor organization, and tools (Tapia & Banegas 1990). However, the key difference between my estimates of labor requirements on raised fields and Erickson's estimates of labor requirements (as well as most other models of production on raised fields) is based on the supposition that raised fields cannot be continuously cropped.

For example, Erickson (1993) reports on controlled experiments in several communities in the Peruvian Lake Titicaca Basin, in which the labor for construction of one hectare of raised fields varied considerably. Erickson estimates between 200 and 1,000 person-days[4] of labor per hectare of raised fields. In comparison, Tapia and Banegas (1990:97) calculated labor requirements based on work in 10 communities in the department of Puno, Peru. They calculated that it took approximately 760 person-days to rehabilitate one hectare of raised fields with adjacent canals.[5]

Based on the hypothesis that raised fields can be continuously cropped, Erickson (1993) contends that raised fields are not labor intensive in the long-term since they require minimal maintenance for up-keep of the system. Erickson estimates that when the total labor for construction and cultivation on raised fields is spread out over a period of 10 years, the fields would only require an average of 270 person-days per hectare per year for construction, maintenance, and cultivation. The problem with this model is that Erickson does not back up his estimates of labor requirements over a 10-year period, with equivalent data on 10 years of production on raised fields. As I have argued in Chapter 5, researchers have not actually demonstrated that raised fields are capable of maintaining a high yield over a long period of time through continuous cropping. Erickson's model is based on the assumption that raised fields are continuously cultivated with potatoes. This was simply not the case for the raised fields in the Bolivian development project, which saw significant declines by the second and third year of cultivation.

Bolivian raised fields, both in Wankollo and in other lake basin communities across the Tiahuanaco and Catari Valleys, were productive for only 2 to 4 years. Therefore, the large initial labor investment calculated by Erickson must be absorbed by a much smaller time frame than the 10–year time frame of Erickson's analysis. Regardless of whether their abandonment was due to a decline in production as Wankollo residents assert, or due to a "cultural resistance" to continuous cropping (Erickson and Brinkmeier 1991; Kolata et al. 1996), no raised field in Wankollo was cultivated in potatoes for more than 4 years.

This is a general trend that was observable across the Bolivian altiplano, since the majority of raised fields associated with the NGO were built in 1990 and 1991. Based on my regional survey of 12 communities in 1994 with a team of researchers hired by the Inter-American Foundation, we saw that many raised fields were beginning to be abandoned. By my subsequent return to Bolivia in 1996, all of the raised fields had been abandoned. As I drove across the altiplano in November 1996 and again in March 1997 to revisit the communities with raised fields from my 1994 field season, I did not see any of the fields that had been in production in 1994 still in production by 1996–97.

I argue that regardless of their purported potential according to development models, and regardless of whether that failure was due to social, technical, or political factors, the fact is that the raised fields built by the NGO in Bolivia were abandoned within four years. None of the raised fields in any community participating in the project for the NGO *Fundación Wiñaymarka* were cultivated for 10 years, as Erickson's model of raised fields states. Thus a labor investment model based on the raised fields being cultivated continuously with potatoes for 10 years is not an appropriate model for the Bolivian raised field case. Instead, I argue that if labor investment is averaged into a four-year production model on raised fields, the maximum number of years that the fields were actually cultivated in Wankollo and in other communities, then the high initial investment simply does not make economic sense.

Unfortunately, from both the experiments by archaeologists and the large-scale building of raised fields by the NGO, there is no published data on labor estimates for raised field construction and cultivation in the Bolivian Lake Basin. For example, Kolata et al. (1996) do not consider the cost of labor in their raised field production model and so this variable was neither measured nor commented on. As I have noted previously, Kolata et al. (1996) also claim that raised fields are capable of continuous cropping (which is implicit in their model of productivity) without actually providing long-term data to support this claim. Thus, I rely on the data of researchers on the Peruvian side of the lake, who both documented labor and calculated labor estimates in the construction and cultivation of raised fields.

Now that we have estimates of labor for construction of raised fields, let us compare this with regular dry fields, which is a more extensive use of the land. Regardless of whether the construction of one hectare of raised fields took 200, 750, or 1,000 person-days to complete, how does this compare to the amount of labor invested in preparing one hectare of regular fields? And further, how does the labor on regular fields and on raised fields compare to the relative production on each type of field? I will argue that it is far simpler and much less labor intensive to cultivate a regular dry potato field than it is to build and cultivate raised fields using only manual labor. This is not surprising since regular dry field farming is certainly a much more extensive use of the land than the highly intensive raised fields. Further, I argue that raised fields require too large an investment of time and energy, when such labor is scarce and perhaps could better be applied towards wage labor. I also make the argument that though production per unit of land was higher on raised fields, the production per unit of labor was not.[6]

In comparison to the estimates that ranged widely from between 200 and 1,000 person-days to build one hectare of raised fields, regular dry fields are much more simply and quickly prepared for cultivation. This is particularly true in Wankollo, since most families paid to have their fields plowed with a tractor and used oxen to further prepare fields for cultivation. If families paid to have their new potato fields plowed, then preparation time and labor is drastically reduced. After the fields were plowed, farmers usually go over the field a few times with oxen-drawn plows to break down large chucks of soil left by the tractor and to aerate the soil. However, one man with a pair of oxen could easily do this work in a day. If the field was not first plowed by tractor, then plowing it by oxen took a bit more time. The farmer would return to the field a few times, between two to 10 days depending on the size of the field. He would first plow the field in one direction such as north to south, then plow the field in the other direction east to west. Though there were few actual fields over ½ hectare in size, the cost of plowing and preparing one hectare with a pair oxen was about 10 person-days.[7] *Wachus,*[8] or lazy beds, were not built in Wankollo and I did not see them very often in other nearby communities, though they were frequently used on the Copacabana peninsula.

After preparing soil for cultivation, it takes a few more days to transport dung fertilizers to the field depending on the field size and how far the field was from the home. Planting the field takes an additional one to two days. During the planting, people work in a team of three with one man plowing, followed by someone dropping the seed in the furrow, followed by a person who distributed dung into the furrow. One of the party would then go back and cover over the newly planted potato field. The planting of one hectare of potato fields probably took no more than 12 person-days, since a group of three could easily plant a ¼ hectare field in a day.

Labor requirements for harvesting both raised fields and regular fields are extremely variable, and it is directly in proportion to the production on the field and the crop yield. A particularly good harvest on any field will take much more labor time than a poor harvest on a equivalently sized field. For example, one farmer said it took her 5 days working alone (5 person-days) to harvest 5 cargas from her ¼ hectare field of potatoes, because she had lost much of the field to worms (*gusano*) and frost (*helada*). While for another farmer, it took him 6 days working with his wife (12 person-days) to harvest 15 cargas from his ¼ hectare field.

Therefore, it is evident that raised fields take many more person-days to build and prepare for cultivation since this labor is manual and very rigorous. But how does production per unit of labor on raised fields compare to regular fields? Rather than project production and labor over a 10-

year time span as Erickson did (1988a), I will compare it over a four-year time span, which was the maximum number of years that any of the fields were cultivated in Wankollo. A four-year time span for the raised fields also represents a general trend in the limits of cultivation on the raised fields built by the NGO *Fundación Wiñaymarka* in the Bolivian Lake Titicaca Basin. In fact, most fields built by the NGO were cultivated for only two to three years.

In contrast, regular dry fields are usually prepared from soil that had been in fallow for many years and the production cycle for potatoes is limited to a single crop per field. In Wankollo, potato fields are planted for one year and followed by a rotation of other crops such as *habas* or *quinua* in the second year, followed by barley in the third year, and ending in a period of fallow. This is a similar pattern found by Carter (1964) and Birbuet (1992), who conducted rural agricultural research in different Aymara communities on the northern Bolivian altiplano. Occasionally in Wankollo, when soils are considered to be exceptionally good, a farmer may decide to plant a second year of potatoes. Of course, the farmer expects production to decrease dramatically in the second year. However, if the soil is of high enough quality and able to produce a second crop (albeit a much reduced second crop of potatoes), he or she may deem it worthy of the price of seed and the small amount of labor required to plant a potato field for a second year. However, I base my labor estimates on fields that are only planted a single year with potatoes.

According to Erickson (1993) raised fields take between 200 to 1,000 person-days per hectare to build and prepare for cultivation. However, when put into the long-term, raised fields average about 50 to 250 person-days per hectare per year, if the initial cost of building the raised fields is divided by four (the maximum number of years for production found in Bolivia). In contrast, in Wankollo it took between 2 to 13 person-days to prepare one hectare of regular dry fields for planting with the aid of a tractor, and between 12 to 24 person-days when using an oxen-drawn plow to prepare the fields. The variability in preparation time had to do with the distance of the field from the home, and the amount time necessary to transport fertilizers (dung) to the field. In the following estimates, I will use the higher estimate of 13 days with tractor and 24 days with oxen-drawn plow. Since regular dry fields usually cannot be replanted a second time in potatoes, these estimates are the same for each year. Therefore, when the cultivation of raised fields is averaged into a four-year time span, they required between two to ten times as much labor as a regular dry field cultivated by oxen plow. When regular dry fields are prepared with the aid of a tractor, the relative labor costs on raised fields increase to between 4 and 20 times that of regular dry fields.

It is clear that raised field require much more labor to build and prepare, but how do raised fields and regular dry fields compare according to their production per unit of labor? Erickson writes, "production data from several years of experimental raised field potato harvest range from 8 to 16 metric tons per hectare, or 2–3 times that of the regular potato farming" (1993:407). Kolata et al. (1996:227) estimate "on 12 cultivated parcels of experimental raised fields in the 1987–88 growing season, potato yields reached an average of 21 mt/ha, or nearly twice the yield of traditional fields treated with chemical fertilizers and over seven times the yield of unimproved traditional cultivation." However, in the 1990–91 agricultural season in 15 communities that actually participated in the NGO project, the average net yield was only 14.85 metric tons per hectare, with considerable variability between 3.6 to 38.5 metric tons per hectare (Kolata et al. 1996) (see Table 6—Potato Production on Raised Fields).

Table 6: Potato Production on Raised Fields		
Experimental Raised Fields in Huatta, Peru, 1981-83[9]	Experimental Raised Fields in Bolivia, 1987-88[10]	Raised Fields built by NGO in Bolivia, 1990-91
10.6 mt/ha	21 mt/ha	14.85 mt/ha

Table 7: Potato Production on Regular Dry Fields		
Potato Production Lake Titicaca Basin, Peru, 1972[11]	Potato Production, Katari Valley, Bolivia, 1987-88[12]	Potato Production in Wankollo, Bolivia, 1996–97
1.21 mt/ha	2.43 mt/ha	3.3 mt/ha

In Wankollo, production on raised fields was good during the first year, relative to regular dry fields, though not as spectacular as some communities surveyed by Kolata et al. (1996). According to production data in Kolata et al. (1996:216), Wankollo had only average results during 1990–91 agricultural season, relative to the other communities evaluated in the survey, yielding between 10.9 to 14.9 metric tons per hectare of raised fields. In comparison, during the 1996–97 agricultural season regular dry potato fields averaged 3.3 metric tons per hectare in Wankollo (see Table 7: Potato Production on Regular Dry Fields and Table 8—Potato Production in Wankollo, 1996–97).[13]

Table 8: Potato Production in Wankollo, Bolivia, 1996–1997

Total Hectares of Potatoes Cultivated	34.63 hectares
Total Cargas of Potatoes Harvested	1,585 cargas[15]
Total Kilograms of Potatoes Harvested	114,120 kg
Average of Cultivated Fields per Household	0.693 ha/household
Average Cargas per Hectare	45.77 cargas/ha (3295.88 kg/ha)
Average Kilograms per Hectare	3295.88 kg/ha (3.3 mt/ha)

However, the problem with both sets of estimates (Erickson 1993; Kolata et al. 1996) is that there has not been systematic long-term data collection on production from raised fields. Therefore, though both Erickson (1993) and Kolata (1996) estimate that in the first year of production raised fields averaged anywhere from two to seven times more potatoes than regular dry fields, this most certainly fell—and fell dramatically—in the second, third, and forth years. Yet even when put into a one-year time frame, for which we do have adequate data to make an argument, we find that though raised fields produced two to seven times as much potatoes per hectare, the labor investment in building one hectare of raised fields was between 8 to 40 times as much as that spent preparing a hectare of regular dry fields with the use of oxen. With the addition of a tractor, the comparison in labor saved is overwhelmingly in favor of the regular dry fields. Thus this so-called "patently ineffective" method of dry field farming currently in use in the Lake Titicaca Basin, when seen in light of production per unit of labor rather than fixating on production per unit of land, is arguably a more rational and labor-cost effective method over raised field farming in contemporary Bolivian communities.

For example, see Table 9—Returns on Labor in Potato Field Preparation. In this table, I demonstrate that when you compare the labor estimates for preparation of regular dry fields prepared by tractor (the most common way of preparing potato fields in Wankollo in 1996–97), with labor estimates for constructing raised fields, the returns to labor per person-day are significantly higher on the regular fields (220 kg) vs. the raised fields (74.25 kg). Even when the preparation time is adjusted for dry fields that are prepared only using oxen drawn plows, we see that the returns to labor are still higher on regular dry fields (137.5 kg). Likewise, even with the phenomenal production yields of 21 mt/ha reported on experimental raised fields by Kolata et al. (1996), the returns to labor on raised fields still only raises to 105 kg per person-day.

Table 9: Returns on Labor in Potato Field Preparation

	Regular Dry Fields prepared by tractor, Wankollo, 1996–97	Regular Dry Fields prepared by oxen, Wankollo, 1996–97	Raised Fields built by manua l labor by NGO, Bolivia, 1990–91[16]
Labor Estimates for Construction and Preparation of Fields person-days/hectare days/ha[18]	13 person-days/ha[17]	24 person-days/ha	200 person-
production/hectare	3.3 mt/ha	3.3 mt/ha	14.85 mt/ha
returns/person-day	220 kg	137.5 kg	74.25 kg

CONCLUSIONS

The problem with previous models of production on raised fields is that they have focused primarily on production per unit of land at the expense of other economic indicators. This has resulted in research that is fixated on proving that raised fields produce much more than regular dry fields per unit of land. Such a focus on production per unit of land is more relevant in rural economies where access to land is the major limiting factor for increasing smallholder agricultural production. Yet I have demonstrated that land is no longer the dominant issue in smallholder agricultural production on the northern Bolivian altiplano due to a variety of factors, most important among these is the large-scale and permanent out-migration of residents, as well as off-farm wage labor employment by male members of the household.

Instead, I argue that the most important factor in the raised field equation is production *per unit of labor*. Erickson's model of production on raised fields does not adequately evaluate this factor on raised fields, since he bases his model on the assumption that raised fields can be continuously cropped in potatoes for 10 years. In the Bolivian Lake Titicaca Basin raised fields were not continuously cropped for 10 years, thus his calculations of labor investment over the long-term are not accurate in this case. If adjusted to a four-year cropping cycle on raised fields, the labor investment on the raised fields is significantly higher than the labor invested in four years of regular dry farming. Since systematic production records on the continuous cropping of raised fields have not been kept, it is difficult to assess the labor investment of the fields over the long-term relative to their long-term productivity. However, it is certain that declines in the initial high yields obtained by both Erickson and Kolata would happen during the second, third and forth years of production in potatoes on raised fields. In fact, this is exactly what happened in Wankollo and the ultimate reason why all the fields were completely abandoned after 4 years.

Based on the production model in Kolata et al. (1996), the authors neither measured nor accounted for the amount of labor necessary for the construction of raised fields in Bolivia. Further, the NGO *Fundación Wiñaymarka* did not adequately address this issue in their development plan for the long-term rehabilitation of raised fields after the inevitable decreases in production following the first season and the withdrawal of development aid, organization, and incentives. Since I argue that labor is the most important production factor for contemporary smallholder agri-culture on the altiplano, endless estimates of production per hectare on raised fields vs. regular dry fields are useless unless they are compared with estimates on the quantity of labor that is invested to obtain these high pro-duction yields.

It seems that researchers have once again fallen for the "myth of the idle peasant" that Stephen Brush argued against for highland Peru over twenty years ago. There is not an endless supply of labor in the countryside and never has been. Today in Wankollo, most children in any given family are migrating out of the community permanently when they are adults in order to seek wage work. The families that remain do not meet their own subsistence needs through farming, and nearly all families are forced to purchase potatoes for consumption after the spring planting. Most families are dependent on some form of off-farm employment or off-farm income. They need money to pay for their children's educational needs and to give them a better life. They are not going to divert their precious time building and cultivating a system of subsistence agriculture based on backbreaking manual labor. Farmers are not interesting in merely increasing production per unit of land at the expense of many days of hard physical labor, not when this labor might be better spent working on other projects and seek-ing off-farm employment.

NOTES

[1] This argument is based upon the premise that raised fields are not continuously cultivated, regardless of whether they may have had the capacity to do so or not.

[2] Two criteria for community "membership" are the simultaneous ownership of a *sayaña* accompanied by active participation in the *sindicato*. There are individuals who own land in Wankollo who do not participate in the community *per se* and

thus are not considered full community members by other residents. Likewise, guardians (cuidadores), who live in the community and take care of property for persons who live outside of the community, are not considered community members.

[3] For further comparison, people working on archaeological excavations during the 1996-97 seasons received between 25 and 35 bs. per day depending on experience and whether they were members of the local archaeology workers union. I paid my own research assistant 30 bs. per day.

[4] A person-day equaled 5 hours for strenuous tasks such as building and cultivating raised fields. I chose to make all participants equivalent units so that my labor estimates are comparable with data from Erickson (1985,1988b), while Mathewson (1987) estimates a 6-hour person-day for building raised fields in the Guayas region of Ecuador. Regrettably, this is not the most sophisticated system of analysis since it equates both genders and all age categories as equal (Sanabria 1993:153).

[5] However, Tapia and Banegas base their estimate on a ratio of planting surface to canal equal to 1:1, with the field surface equaling about 50% of total land. Kolata et al. (1996:213) report that the cultivatable surface for each hectare of raised fields in the Bolivian Lake Titicaca Basin was about 60%.

[6] I compare estimates of person-days (based on a five-hour day) for building and cultivating raised fields with estimates of person-days for preparing and cultivating regular dry fields. I demonstrate that the ratio of production per person day is higher on regular fields than it is on dry fields.

[7] This is based on the estimate that it took one man with a pair of oxen approximately two weeks (10 days) to go over a field enough times to thoroughly aerate and break up the soils.

[8] Wachus are small, mounded furrows for planting potatoes.

[9] Based on Erickson (1988a:245).

[10] Based on Kolata et al. 1996:227).

[11] Based on Golte (1980:113).

[12] Based on Kolata et al. (1996:227).

[13] Households (n=50) estimated production on each potato field that they cultivated that season. The total production (114,120kg) was divided by the number of hectares (34.63 ha) to arrive at the average production per hectare (3.3 kg/ha.).

[14] Potato production figures based on data collected from 50 households in Wankollo.

[15] 1 carga = 72 kg

[16] Based on Kolata et al. (1996).

[17] Average based on a one-year production cycle, with initial opening of fields done by tractor.

[18] Labor estimates based on those obtained by Erickson (1988a).

CHAPTER SEVEN

Conclusion: Inventing Indigenous Knowledge and the Maintenance of Class and Ethnic Boundaries

C ULTURAL TRADITIONS ARE NO LONGER THOUGHT TO BE STATIC collections of customs and beliefs that are handed down from generation to generation (Handler and Linnekin 1984; Hanson 1989; Hobsbawm 1983). Instead, cultural traditions are viewed as dynamic and flexible, often integrating contemporary values and politics, while simultaneously staking a claim in the past. Anthropologists and historians have become more aware of the role that they themselves play in this interpretive process of delegating certain practices and beliefs to be labeled as "traditional" (Handler and Linnekin 1984). Cultural anthropologists, in particular, have grappled with the question of how our ethnographies represent other cultures and other peoples. Thus, one recurrent theme in the interpretive anthropology in the 1980s and 1990s focuses on the "doing and writing of ethnography," rather than on the building of general theories of culture (Marcus and Fischer 1986:16).

This book argues that raised fields are an invented tradition of agriculture; invented by archaeologists and development workers who were interested in rehabilitating a method of agriculture that they (the academics and development workers) deemed to be ancient, ancestral, and natural to the Lake Titicaca Basin. Other works have previously examined the role of ethnography and history in the creating of traditions and the institutionalization of traditional practices (Friedman 1992; Hanson 1989; Hobsbawm and Ranger 1983; Nordholt 1994). However, with the raised field rehabilitation project, this book has the unique perspective of: 1.) critiquing how archaeologists and development workers represent raised fields and local peoples, and 2.) examining the process of institutionalizing

155

(or attempting to institutionalize) the practice of raised fields agriculture through a contemporary rural development project.

The raised field rehabilitation project, which attempted to rebuild and resurrect the ancient raised fields of the pre-Hispanic civilization of Tiwanaku, was a creation of North American and Bolivian archaeologists and development workers. Their aim was to rebuild a system of agriculture based upon their own understanding and interpretation of the pre-Hispanic past in the Lake Titicaca Basin. However, what these researchers did not account for in their development scheme, was their own role in both interpreting the pre-Hispanic past and in designing a rural development project based on their interpretations. I maintain that the creating of any economic development project based on contemporary representations of the past, inherently calls for a critical understanding of the role of contemporary forces in forming these interpretations of the past. In chapter 3, I have deconstructed some of the interconnected social, political, and economic themes about indigenous peoples and sustainable development, which influenced and informed the representations of the raised field rehabilitation project.

I argue that it was the cultural values and political interests of the archaeologists and development workers, which led them to label Lake Titicaca Basin peoples as indigenous, and to cast them in the role of the simple, non-modern, and indigenous agriculturalist. Likewise, it was the aim of archaeologists and development workers that these people cultivate an organic, regenerative, and ecologically sustainable method of agriculture, as they believed raised fields to be. However, this did not interest Wankollo residents, who were willing to experiment with the raised fields given the fields' impressive initial yields and other development incentives, but eventually abandoned them when these incentives where withdraw and the productivity of the fields declined. It was not the ambition of the residents of Wankollo to rebuild and rehabilitate an ancient and indigenous technology recovered from their ancestors.

In chapter 4, I demonstrated how archaeologists and development workers represented raised fields and local residents of the Lake Titicaca Basin as indigenous. The representation of raised fields produced a stereotypical image of a timeless and simple "Indian" farmer through the emphasis on raised fields as appropriate technology. Likewise, representations of raised fields as indigenous knowledge and sustainable development portrayed local peoples as having an innate sense of harmony and understanding of their environment. This representation of indigenous peoples as natural conservationists and innate ecologists was a based on Western preconceptions of indigenous peoples (Krech 1999).

Krech (1999) maintains that Westerners, particularly North Americans of European ancestry,[1] hold a deep-seated preconception of Native Americans as natural conservationists and proto-ecologists; they were "ecological Indians." Krech (1999:17) argues that the imagery of Native Americans as conservationists has its roots in over two and a half centuries of social history, from when Europeans first produced these images as a way to represent the New World and its inhabitants.

Along with this representation of Native Americans as conservationists and proto-ecologists, recent work by Kempton, Boster, and Hartley (1995) has shown that North Americans have becoming increasingly interested in environmentalism and human impacts on the environment, for example with global warming. The authors write, "environmentalism has already become integrated with core American values such as parental responsibility, obligation to descendants, and traditional religious teachings" (1995:214). Given North American interests in environmental issues, and an internalized image of indigenous peoples as innate environmentalists, this combination creates a dynamic setting for economic development. I would argue that this uncritical association of indigenous people with the environment, and American interest in environmental issues, has driven recent interests in indigenous knowledge as a model of sustainable development.

The raised field project, which sought to rehabilitate an ancient indigenous technology, appealed to Western interests in ecology and the natural environment, especially as it was tied to sustainable development. Researchers and development workers uncritically made the link between sustainable development and indigenous technology, perhaps drawing on their own deeply held preconceptions of indigenous peoples and Native Americans.

Drawing on representations of raised fields as indigenous knowledge, archaeologists and development workers made an uncritical connection between this indigenous knowledge, and the values of North American environmentalism and ecology. In implementing the raised field rehabilitation project, I put forward that archaeologists and development workers created an "invented tradition" of raised field agriculture (Hobsbawm 1983). Through the creation and implementation of the raised field rehabilitation project, and by labeling this project as a form of indigenous knowledge, archaeologists and development workers were inventing indigenous knowledge.

Research by archaeologists, cultural anthropologists, and historians has looked at the ways archaeology reinterprets and creates the past, and how it can be exploited by emerging nationalist, ethnic, and indigenous movements (Benavides 1999; Gathercole and Lowenthal 1990; Schmidt and Patterson 1995). For example, Benavides (1999) examines the role of archaeology in the "historical legitimization" of Ecuador as a nation-state, and how national heritage is constructed through the appropriation of archaeology and archaeological data. This book follows in a similar vein by looking at the ambiguous ways that applied archaeology and a rural development project invented an image of raised fields as indigenous knowledge, and imposed this representation onto local residents in the Lake Titicaca Basin.

But what are the political and social implications of representing raised fields as indigenous knowledge? What would be accomplished by rehabilitating these ancient raised fields and by putting the contemporary cultural representation of raised fields as indigenous into practice through a rural development project? To answer these questions, I turn to a closer look at the power that social representations of indigenous knowledge have over the politics of development, particularly as it is practiced under the sustainable development paradigm.

I begin this discussion by making two general statements about cultural representations. 1) Dominant cultural representations define subordinate groups and define how people in society think and relate to them. 2) Cultural representations define the boundaries of discourse, thereby maintaining social boundaries between groups of people, such as boundaries of ethnic, class, sexual orientation, or gender, among others (Escobar 1995; Kearney 1996; Said 1979).

For example, Kearney (1996) examines how anthropologists have represented "peasants" by concentrating on a set of social, economic, and political characteristics to define and describe these peoples who have been the subject of much anthropological inquiry. Anthropological representations of peasants have created an essentialized image of peasants and the peasantry. Additionally, Kearney contends that there is a problematic concerning these essentialized anthropological representations of peasants:

> To name, that is, to label persons or an entire community as a certain type and then to elaborate a theory of their essential social identity is to create a symbolic representation of those persons or that community. If this representation then becomes naturalized, that is accepted as the "obvious" depiction of its referent, it becomes a mold that shapes the second sense of representation—the political (Kearney 1996:171).

Now if we relate this argument to the raised field project, we see that archaeologists and development workers labeled the inhabitants of the Lake Titicaca Basin as indigenous, and with this label came the requisite characteristics (in certain respects similar to peasants), that they are simple agriculturalists in harmony with their natural environment. I would argue, like Kearney's representation of peasants, the representation of raised fields and the peoples of the Lake Titicaca Basin as indigenous, was accepted as natural or "obvious," at least by the archaeologists and development workers who created and implemented the project.

Escobar's (1995) work also analyzed representations, though his subject focused on the representations of the "Third World." Escobar uses the concept of a "development discourse" to analyze how the Third World was produced and represented by development agencies and development projects around the world. For Escobar, by approaching the subject of development as a discourse, it made possible an analysis of the mechanisms of hegemonic power that maintains and perpetuates certain cultural representations over others. Escobar's work on the discourse of development made possible a far more thorough analysis of social representations of the Third World created and perpetuated by Western economic development agencies. Escobar's work is a detailed study of the ways that Western development continues to manage and reproduce the Third World throughout the latter half of the 20th century.

However, the current case study of raised fields in contemporary practice in Bolivia diverges from the global and all encompassing scope of works like Escobar's, which attempt to characterize all development discourse as a single totalizing discourse. This book studied one specific rural development project and focused on how global academic and development discourses about indigenous peoples influenced the acceptance, implementation, and representation of the project. Specifically, this book focused on the discourse of raised field rehabilitation in Bolivia.

Ferguson's (1994) ethnography on Lesotho also studied the implementation and aftermath of a "failed" development project. For example, Ferguson describes how the livestock production plan in Lesotho was never accepted and nor implemented in the way that the planners had imagined it to be (1994:20). As I have argued, the discourse that represented the raised fields project as indigenous, pre-Hispanic, and sustainable, was never accepted by those who would be reconstructing and cultivating the raised fields—the Aymara speaking residents of the Bolivian Lake Titicaca Basin, such as those in Wankollo. Thus, it is not surprising that local inhabitants in each case eventually abandoned both the Livestock Production plan in Lesotho studied by Ferguson, and the raised field rehabilitation project in Bolivia.

In conclusion, how has the representation of raise fields as indigenous knowledge defined the local people of the Lake Titicaca Basin? How does the invention of indigenous knowledge actually impact the lives of Bolivian Lake Titicaca Basin inhabitants? Raised fields and the raised field rehabilitation project utilized representations of indigenous peoples and indigenous knowledge that depicted local residents as subsistence farmers. It emphasized their indigenous origin, which linked these people to a pre-Hispanic past.

The raised field rehabilitation project made a seamless connection between current residents of the Lake Titicaca Basin, and the peoples who cultivated the ancient raised fields of Tiwanaku over 1,000 years ago. I suggest that the raised field rehabilitation project was maintaining and reproducing the dominant discourse about indigenous peoples and peasant farmers. This discourse represented them as simple, traditional, rural, natural, and aboriginal, all of which are oppositional to how Westerners, such as the archaeologists and development workers, are likely to represent themselves. Thus the question is, to what purpose does such an oppositional discourse serve? What affect does a development project that perpetuates these representations have on the people it is representing?

Implementing an invented tradition of raised field on a large scale would have kept local farmers working in the countryside doing subsistence agriculture in an essentialized and timeless fashion. Raised field agriculture keeps rural "Indians" working on their farms and maintains their "Indianess," particularly as it is viewed in connection to the land, to pre-Hispanic history, and to the idealized image of indigenous peoples as proto-ecologists. The raised fields "contained" the indigenous Aymara speaking farmers by keeping them on their farms and in a defined subordinate position as peasant agriculturalists (Kearney 1996).

The raised field project also maintained the boundaries of ethnicity and class as defined by the dominant Bolivian and North American cultures. Let us take for example, an analysis of the academic representations of peasants and the peasantry, and how this applies to the case of raised fields. The seminal work by Wolf (1955) defines "types of Latin American peasantry" as people who have their primary economic livelihood focused on agricultural production, who have security of tenure to land, and whose household economies are oriented towards subsistence rather than markets and reinvestment. As Kearney notes, this image of the Latin American peasant became reified in anthropological works as a social category that "more often than not has no singular objective ethnographic or class basis" (1996:2).

Turning now to the case of raised fields, I would hold that we see a similar reified image of indigenous peoples in the representation produced by the archaeologists and development workers. Local peoples are represented as primarily involved in agriculture and oriented towards subsistence. As I have argued in chapter 6, it was the assumptions about the availability of labor based on preconceptions of indigenous agriculturalists, which proved to be the primary factor in the contemporary abandonment of the raised field project. But by returning local residents to subsistence farming in the anthropological image of the subsistence peasant farmer, these local peoples are kept working in the countryside, perhaps where they belonged according to the wishes of urban Bolivians and the romantic notions of North American researchers.

When residents of the Lake Titicaca Basin abandoned the raised fields and returned the fields to fallow after the withdrawal of development aid, perhaps they were sending their own message and writing their own counter-discourse. Through their actions, Wankollo residents were saying that they did not see themselves in the role of indigenous subsistence farmer, cut off from the economy and politics of the nation. One could view the "failure" and abandonment of the 20th century raised field rehabilitation project as a symbolic act. Local residents were abandoning and discarding the symbolic domination and subordination that the practice of raised fields would have maintained rather than alleviated.

In Wankollo, all households had direct connections to wage labor opportunities or other means of generating a cash income. Often this included some members of the immediate family living and working away from the community and their farms. Families in Wankollo were already participating in the national economy and engaged in local, regional, and national politics, as they have been since well before the National Revolution as demonstrated in Chapter 2. Is this the portrait of a timeless indigenous farmer who is ecologically in tune with nature, apolitical, and practicing subsistence agriculture? Though archaeologists and development workers may have represented rural farmers in the Lake Titicaca Basin as subsistence cultivators having abundant time and labor to invest in intensive forms of agriculture, the residents of Wankollo certainly did not internalize this representation. It is quite clear that Wankollo residents discarded this representation of themselves as simple indigenous farmers, when they abandoned raised fields.

I conclude this book with a story of two little girls in Wankollo, one 8 years old and the other 10 years old. The older of the two was the daughter of community members who cultivated their lands and sought wage work when it was available. The second little girl was the granddaughter of a prominent family, whose father was a mechanic in La Paz, and whose uncle was a lawyer. Both her father and uncle participated fully in the social and political life of Wankollo, had land in the community, and returned often on the weekends and for the monthly community meetings, and other activities. I asked these two girls the typical North American question of, "What do you want to do when you grow up?" The older of the two whose parents were full-time farmers, said she wished to own her own shop in the city and be a storekeeper. While the younger of the two who lived in the city with her father and her uncle, said to me that she wanted to be a lawyer like her Uncle Antonio.

The fact that my two little friends could imagine their lives as something other than subsistence farming demonstrates that while archaeologists and development workers might represent local peoples in a timeless, traditional, and indigenous fashion, these representations have not necessarily been internalized. For my two young girlfriends, they were already imagining a different future for themselves.

NOTE

[1] Krech (1999) focused primarily on North Americans, though I would add that his argument could also be extended to South Americans of European and *mestizo* descent.

Methodology

I ARRIVED IN BOLIVIA IN MAY OF 1994 TO CARRY OUT A PILOT STUDY ON the economics of raised field agriculture in the Bolivian Lake Titicaca Basin. My first goal was to make contact with persons working on various raised field projects in Bolivia. I was first introduced to the director of the National Institute of Archaeology and Anthropology who was also the head of the NGO *Fundación Wiñaymarka* and had been involved with the earliest experiments on raised fields in Bolivia in collaboration with the archaeologist Alan Kolata of the University of Chicago. I spent many afternoons in the La Paz offices of the NGO *Fundación Wiñaymarka* during my first weeks in Bolivia, where I talked with the coordinator, agronomists, and various staff at the NGO.

I also interviewed other individuals who were conducting or had at one time conducted research on raised fields, as well as individuals who were involved with or had been involved in promoting the rebuilding of raised fields in lake basin communities in Bolivia. They included: the director and one field assistant from *ProSuko*, another NGO interested in researching and promoting raised fields; advanced students conducting agricultural research at the Tiahuanaco field station for the Catholic University in La Paz; the director and assistant director of *Semilla*, an NGO that often worked for the U.S. based Inter-American Foundation and had been contracted to monitor the progress of the raised field rehabilitation project being carried out by *Fundación Wiñaymarka*; and various member of the NGO *Ayni Tambo*, an NGO sub-contracted by Semilla to conduct the actual survey of communities that had participated in the raised field rehabilitation project and to write a monitoring report based on their observations and findings.

This monitoring report was ultimately being provided to and paid for by the Inter-American Foundation, which had funded part of the research and development carried out by the NGO *Fundación Wiñaymarka*. Since I had also been funded by the Inter-American Foundation to carry out master's level research on the raised field rehabilitation project, I was able to join the group. In June and early July of 1994, I joined the team of five researchers from *Ayni Tambo* on a series of field trips to a dozen communities in the Bolivian Lake Titicaca Basin. The group split up into pairs and interviewed several members of each community apiece.

Based on this two-month reconnaissance trip to Bolivia in 1994, I gained some indication that the raised fields were not as successful in common practice as earlier experiments, reports, and publications had portrayed them. For example, in the community of Lukurmata the raised fields had all ready been abandoned. In other communities, residents said that they were not planning to build any more raised fields. However, it was not until after I arrived in Bolivia in March of 1996 that I realized that the NGO *Fundación Wiñaymarka* had completed disbanded, that no new fields were being built, nor were any of the previously built fields still being cultivated. Since my original research design called for detailed qualitative and quantitative interviews that compared contemporary raised fields with regular dry flatland and hillside fields, I was compelled to rethink my fieldwork methodology. The key problem that I faced was how to collect the data I needed on raised field cultivation in modern practice when the fields had ceased being cultivated by the time of my arrival in 1996. I resolved to do this by combining a number of data collection methods including: quantitative economic research on agricultural production in Wankollo, oral histories about raised field cultivation and production with community members who had participated in the project, archival research on documents pertaining to the raised fields, and interviews with development personnel who had worked with the raised field rehabilitation project.

I spent the first four months of my second trip to Bolivia in 1996 reacquainting myself with previous contacts, conducting more interviews with these and other individuals, and studying language (both Spanish and Aymara). I spent many afternoons in those first fours months meeting with various persons at the National Institute of Archaeology and Anthropology, meeting the new director and sub-directors, and other individuals who worked at the Institute.

I also began efforts to select an appropriate research site in which to gather data on the economic conditions for agriculture in the Lake Titicaca Basin and the contexts of raised field rehabilitation in modern practice. Choosing and gaining approval to conduct research in a Lake Titicaca Basin community proved to be my most difficult obstacle. I needed an official sponsor in Bolivia before any community would allow me to do research. It seemed natural as an anthropologist that this sponsor should be the National Institute of Archaeology and Anthropology, who agreed to do this for me and signed an official agreement (*convenio*) giving me permission to work at the site of Tiwanaku. The people of the Tiahuanaco and Catari Valleys had much experience with archaeologists so I was easily associated with them as researchers. Besides, I much preferred to be affiliated with these academic researchers, rather than with the various development groups that had worked in the area.

Bolivia is the second poorest country in the Western hemisphere, thus it should be no surprise that archaeology in the Bolivian Lake Titicaca Basin is dominated by projects sponsored by foreign institutions and researchers. During the course of my own fieldwork (March of 1996 to November 1997), researchers from North American institutions sponsored and directed all of the major excavations taking place in the region. These North American archaeology projects are usually well funded in comparison to Bolivian projects and they are required to hire Bolivian archaeologists, technicians, and other local people to work for them. However, this employment is sporadic and usually seasonal, since most North American archaeologists work during the months of June through September. This fortuitously coincides with the winter months when rural households have more time and labor available to them.

When North American archaeologists returned to Bolivia in June and July of 1996 to commence another season of fieldwork, I joined a small team of researchers sponsored by the University of Chicago when they headed into the field that July. My intent was to introduce myself to the community that they were working in, which had also participated in the raised field rehabilitation project, and to request permission to conduct my research on the community. Unfortunately, this situation did not work out and when the archaeology team left the community after a month and a half of excavations, I went with them. However, these nearly two months spent in the daily company of archaeologists on an actual archaeology project, and the following visits during the course of my fieldwork to numerous other archaeology projects, allowed me a firsthand vantage point on the politics of conducting research in Bolivian lake basin communities. It also gave me an opportunity to talk informally about the expectations and experiences of the archaeologists, both with their past experiments in raised fields and with their current excavations.

I eventually chose to work in the community Wankollo, and with assistance from the Institute of Archeology and Anthropology, I was able to get an introduction to the community. The selection of Wankollo was based on a number of criteria: 1) the community had built a relatively large number of raised fields and some rehabilitated pre-Hispanic terraces; 2) it is a large community with both community built raised fields and a number of individual/private family built fields; 3) the proximity of the town of Tiahuanaco and site of Tiwanaku made it possible to also observe the activities of archaeologists and tourists; and 4) the diverse agricultural and pastoral economy allowed me to collect data on the interface between raised fields and negotiations of resources between different livelihood strategies.

By October of 1996, Wankollo's *sindicato* had met and approved of my research in the community. At that time I was staying in a room at the Tiahuanaco Museum, located on the northwest border of the community on a small road between the town of Tiahuanaco and the center plaza of Wankollo. I remained in the Tiwanaku Museum throughout the period of fieldwork, which allowed me to take part in the community life of Wankollo, the town of Tiwanaku, as well as affording me the chance to talk to archaeologists and tourists who came to visit the site of Tiwanaku.

I began work in Wankollo in October of 1996 with the assistance of Ernesto, the youngest son of a once powerful *ex-comunario* and brother-in-law of the current *secretario general*. Ernesto had been assigned as my assistant and I had little choice in this matter, though he was fluent in Spanish and had worked on other development projects so households were used to dealing with him as liaison with various development projects. Many of the residents of Wankollo always remained suspicious of my intentions and avoided my questions. However, with Ernesto's help I was able to talk to most households in the community. I also established fictive kin ties with the *secretario general,* which helped to convince others in the community to work with me.

My fieldwork in Wankollo was primarily carried out from October of 1996 through September of 1997. Fieldwork consisted of informal interviews and participant observation in a variety of informal settings, as well as semi-structured interviews using interview guides and an interview schedule. Research began in Wankollo with a detailed census of the 90 households, conducted with a household head in most cases.[1] In addition to standard demographic data, the census provided information on level of education, occupation(s) of residents, migration out of the community, number of family members living away from the community permanently, previous *sindicato* positions held by residents, and ownership of plots of land based on the agrarian reform map.

Data on agricultural production and labor allocation for agriculture in Wankollo for chapters 5 and 6 were collected through two series of semi-structured interviews. One series of interviews pertained to information on the potato sowing, while the second series pertained to the harvest. Since participation in my survey was voluntary (we had emphasized this at *sindicato* meetings), my sampling strategy was opportunistic in so much as I interviewed as many of the 90 households as I could easily persuade to participate. However, my assistant and I made a concerted effort to sample both *ex-comunario* and *ex-colono* households. The first series of interviews took place between November and December of 1996, roughly corresponding with the end of the sowing for that year's crops. I interviewed 60 heads of household on the number of fields they planted in potatoes, the number of days and number persons it took to prepare and plant the fields, the relationship of those who helped them in the planting of fields, and other agricultural questions regarding the potato sowing. The second series of semi-structured interviews took place between April and June of 1997, immediately following the potato harvest, when families finally had the time to sit down and talk with me. These interviews were about twice as detailed as the previous series and were conducted with 50 heads of households in Wankollo. In addition to standard agricultural production questions such as the quantity and quality of the harvest, the crops harvested, and dates, I focused a large portion of the harvest survey on labor data, particularly the potato harvest, which requires the largest amount of labor for any single agricultural task.

In calculating my labor data, I chose to follow the precedent set by other archaeologists working on raised fields so that my data would more closely correspond. This was necessary since by 1996 raised fields had been abandoned, so that in comparing my own labor data on regular dry fields with the data on raised fields I rely directly on other researchers for labor estimates on raised fields. In particular, I rely on the data estimates of Erickson (1988a), who experimented with raised fields on the Peruvian side of Lake Titicaca in the 1980s and who provides the most detailed accounting of how he arrived at his estimates and his methodology (see Erickson 1988a:221–238). Erickson based his quantifications of labor based on a 5-hour workday, and giving equal weight to both males and females as equivalent units, though there are methodological drawbacks to this method.

In between these two series of semi-structured interviews, I collected oral histories from January through March of 1997 with Wankollo res-

idents who had participated in various capacities with the raised field rehabilitation project. These oral histories were designed to recollect and reconstruct the history of the raised field project in the community, such as the
number of years crops were cultivated, who participated in the building
and cultivation of the fields, how the labor was organized, and how and in
what ways was the community assisted in their endeavors by the development project. I relied on *Fundación Wiñaymarka's* own production records
to establish the production on raised fields in Wankollo and compared this
with the recollections of the residents concerning production.

Throughout my 20-month stay in Bolivia from March 1996 to
November of 1997, I collected documents and published materials on
raised fields and the raised field rehabilitation project. The *Museo
Arqeológico de Tiwanaku* had a large room with an exhibit on pre-
Hispanic raised fields and the modern raised field rehabilitation project,
which I viewed countless times, though I was unable to photograph it. I
also spoke with tourists who visited the site and observed the daily pil-
grimages of the tourists to the site. I also went on or accompanied several
tours of the site and recorded the ways that tour guides represented the site
of Tiwanaku and the reconstructed raised fields that were impressively on
display when viewed from the top of the highest pyramid at the site of
Tiwanaku.

NOTE

[1] This household head was either the male or female head of the household, I chose
not to privilege one gender over the other, particularly since women tended to be
more available during the days to speak with me and were arguably equal partners
in household agricultural decision making (S. Hamilton 1998).

Figures

Fig. 1: Wankollo Raised Fields

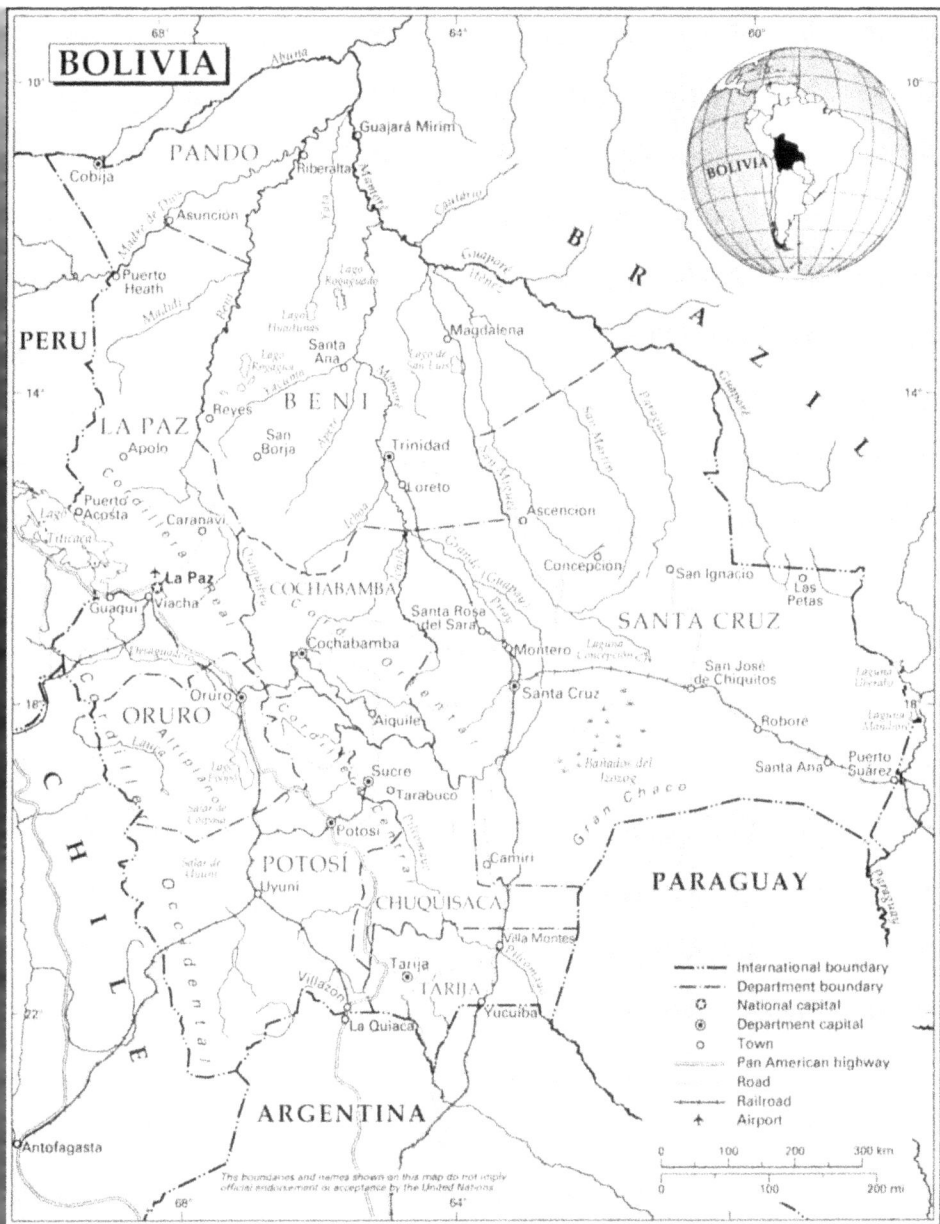

Fig. 2: Map of Bolivia

Fig. 3: Map of the Lake Titicaca Region

Fig. 4: Agrarian Reform Map of Wankollo

*Wali lakaquiwa Tiwanakun ch' iqhi amuyt'
awinacapjj amthapiñasa, acjamarc suma k'uma
jach' ancawi cancañs yatjjatapjjañasaraqui,
amthapipjjañasaraqui.*

*Urge rescatar la sapiencia de Tiwanaku,
pero también los valores morales de sus
creadores...*

Fig. 5: Andean Cross

TIWANAKU

Y

LOS SUKA KOLLU

A 3.850 metros sobre el nivel del mar, muy próxima al lago Titicaca, la urbe prehispánica de Tiwanaku fue la capital de un vasto estado andino.

Tuvo un foco de generación, epicentro cultural, que por la calidad de sus logros, en todos los campos, fue extendiendo su radio de acción hasta desbordar el medio altiplánico. Aún más, su ideología influenció áreas mucho más lejanas.

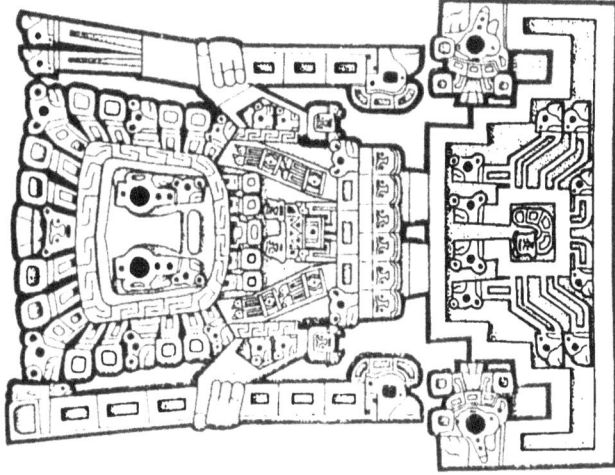

Fig. 6: Staff God

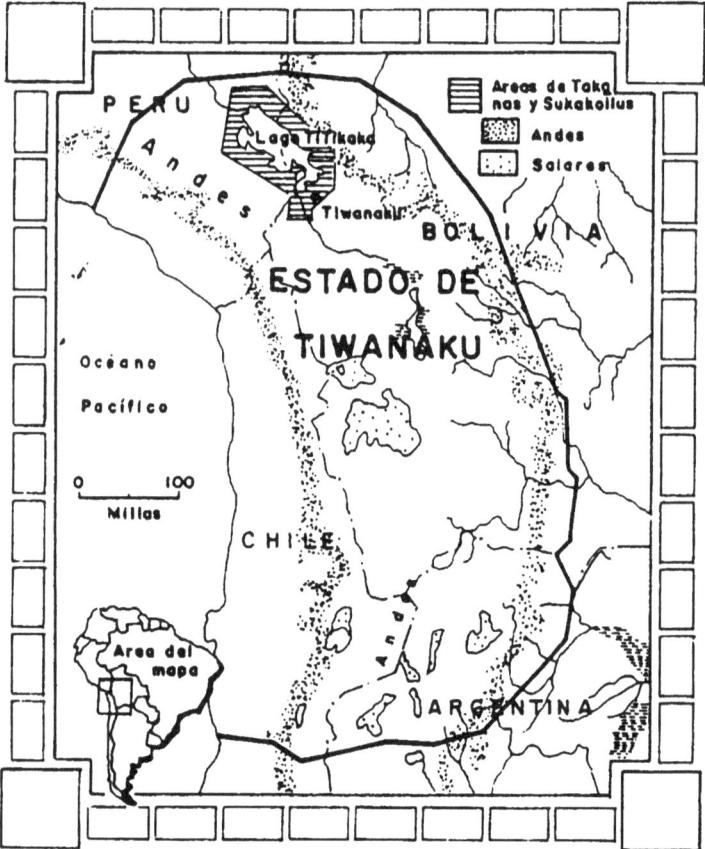

Fig. 7: The Tiwanaku State

Fig. 8: Raised Fields as Organic Agriculture

Fig. 9: Pamphlet Cover

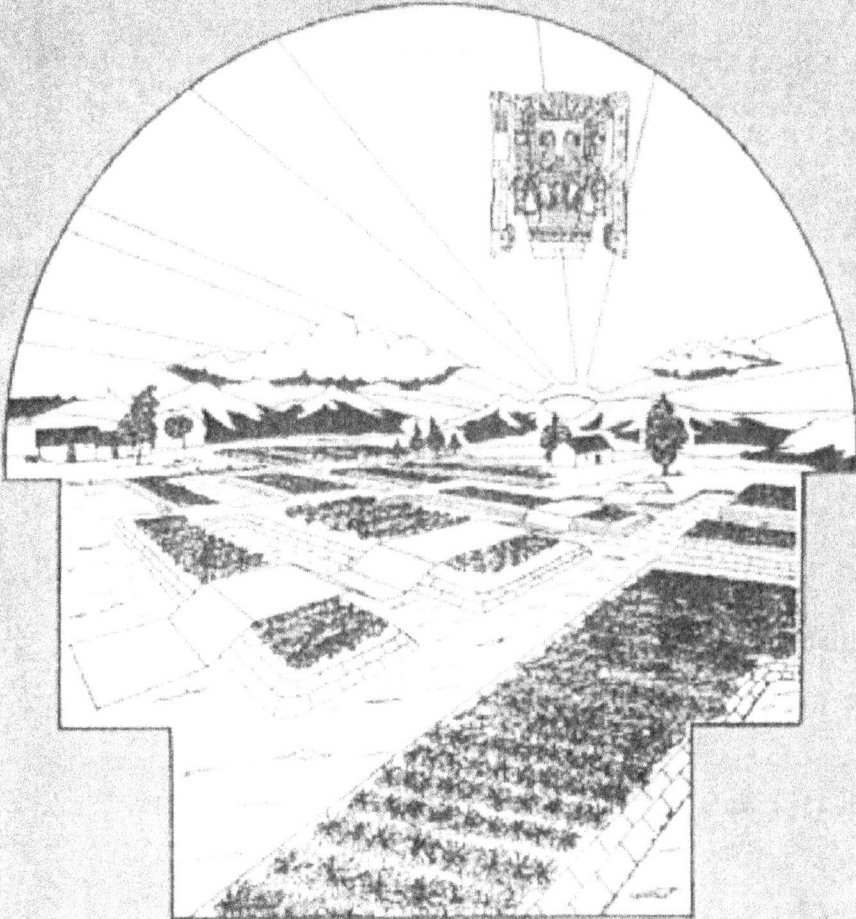

Fig. 10: Pamphlet Cover 2

Fig. 11: Pamphlet Cover 3

¿Qué herramientas se utilizan?

Se utilizan las siguientes herramientas: (si no se las tiene , se las puede hacer)

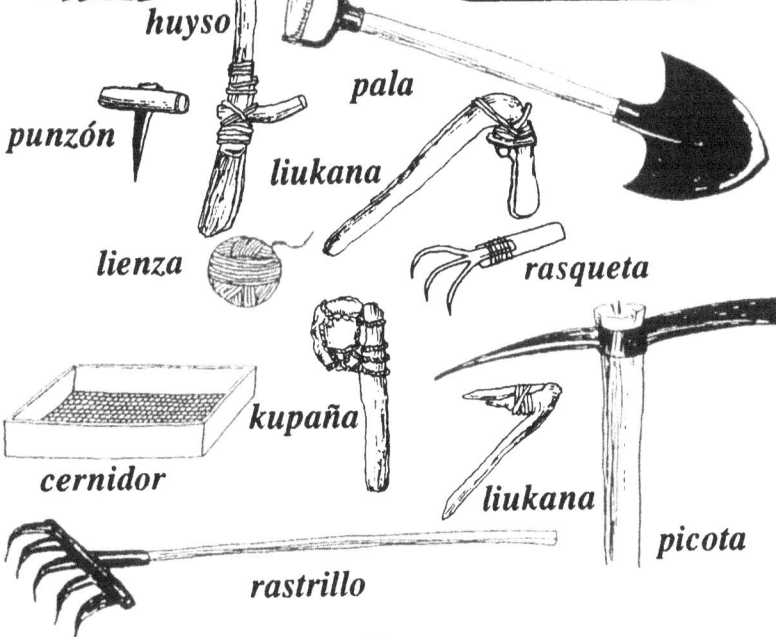

huyso

pala

punzón

liukana

lienza

rasqueta

cernidor

kupaña

liukana

picota

rastrillo

17

Fig. 12: Tools for Raised Field Cultivation

Fig. 13: Representations of Communal Work

Fig. 14: Guaman Poma 1 (*Source:* Guaman Poma de Ayala, Felipe. 1980. El
Primer Nueva Coronica y Buen Gobierno. Translation and textual analysis by
John V. Murra and Rolena Adorno. Mexico: Siglo Veintiuno)

Fig. 15: Guaman Poma 2   (*Source:* Guaman Poma de Ayala, Felipe. 1980. El Primer Nueva Coronica y Buen Gobierno. Translation and textual analysis by John V. Murra and Rolena Adorno. Mexico: Siglo Veintiuno)

Bibliography

Abercrombie, Thomas A.
1997   Pathways of Memory and Power: Ethnography and History among an Andean People. Madison, WI: University of Wisconsin Press.

Abu-Lughod, Lila
1991   Writing Against Culture. *In* Recapturing Anthropology: Working in the Present. Richard Fox, ed. Pp. 137–62. Santa Fe, NM: School of American Research Press.

Adams, William M.
1990   Green Development: Environment and Sustainability in the Third World. New York: Routledge.
1993   Sustainable Development and the Greening of Development Theory. *In* Beyond the Impasse: New Directions in Development Theory. Frans J. Schuurman, ed. Pp. 207–222. New Jersey: Zed Books.
1995   Green Development Theory? Environmentalism and Sustainable Development. *In* Power of Development. Jonathan Crush, ed. Pp. 87–99. New York: Routledge.

Agrawal, Arun
1995   Dismantling the Divide Between Indigenous and Scientific Knowledge. Development and Change 26(3):413–39.

Albó, Xavier
1987   From MNRistas to Kataristas to Katari. *In* Resistance, Rebellion, and Consciousness in the Andean Peasant World, 18th to 20th Centuries. Steve J. Stern, ed. Pp. 379–419. Madison, WI: University of Wisconsin Press.

1994   And from Kataristas to MNRistas? The Surprising and Bold Alliance between Aymaras and Neoliberals in Bolivia. *In* Indigenous Peoples and Democracy in Latin America. Donna Lee Van Cott, ed. Pp. 55–81. New York: St. Martin's Press.

1995   Bolivia: Towards a Plurinational State. *In* Indigenous Perceptions of the Nation-State in Latin America. Lourdes Giordani and Marjorie M. Snipes, eds. Pp. 39–60. Williamsburg, VA: College of William and Mary.

Apffel-Marglin, Frédérique
1996   Introduction: Rationality and the World. *In* Decolonizing Knowledge: From Development to Dialogue. Frédérique Apffel-Marglin and Stephen A. Marglin, eds. Pp. 1–39. Oxford, England: Claredon Press

Armillas, P.
1971   Gardens on Swamps. Science 174:653–661.

Auty, Richard M. and Katrina Brown
1997   Overview of Approaches to Sustainable Development. *In* Approaches to Sustainable Development. Richard M Auty and Katrina Brown, eds. Pp. 3–17. Washington, DC: Pinter.

Bandy, Matthew S.
1999   Productivity and Labor Scheduling Aspects of Titicaca Basin Raised Field Agriculture. Paper presented at the Society of American Archaeology Annual Meeting, March 24–28, Chicago, IL.

Bartlema, Jan
1981   Migraciones Internas Recientes en Bolivia. La Paz: Proyecto INE.

Benavides, O. Hugo
1999   Telling Stories, Producing the Nation: Archaeology's Role in the Construction of Contemporary Ecuador. Ph.D. diss., City University of New York.

2000   Ambivalent Archaeologies and Cultural Production in Ecuador: Narrating, Displaying, and Consuming the Pre-Hispanic Site of Cochasqui. Paper presented at the American Ethnological Society Annual Meeting, March 24–26, Tampa, FL.

Biesboer, David D., Michael W. Binford, and Alan Kolata
1999   Nitrogen Fixation in Soils and Canals of Rehabilitated Raised-Fields of the Bolivian Altiplano. Biotropica 31(2):255–267.

Binford, Michael W., Mark Brenner, and Barbara W. Leyden
1996   Paleoecology and Tiwanaku Ecosystems. *In* Tiwanaku and its Hinterlands. Alan L. Kolata, ed. Pp. 89–108. Washington, DC:

Smithsonian Institution Press.

Binford, Michael W. and Alan L. Kolata
1996 The Natural and Human Setting. *In* Tiwanaku and its Hinterlands. Alan L. Kolata, ed. Pp. 23–56. Washington, DC: Smithsonian Institution Press.

Binford, Michael W., Alan L. Kolata, Mark Brenner, John W. Janusek and Matthew T. Seddon, Mark Abbott, and Jason H. Curtis
1997 Climate Variation and the Rise and Fall of an Andean Civilization. Quaternary Research 47:235–48.

Birbuet Díaz, Gustavo
1992 La Economia Campesina en la Microregion de Caquiaviri y Comanche, Provencia Pacajes. La Paz, Bolivia: HISBOL.

Browman, David L.
1994 Titicaca Basin Archaeolinguistics: Uru, Pukina and Aymara A.D. 750–1450. World Archaeology 26(2):235–251.

Brundtland, Gro Harlem
1987 Our Common Future. World Commission on Environment and Development. New York: Oxford University Press.

Brush, Stephen B.
1977a Mountain, Field, and Family: The Economy and Human Ecology of an Andean Valley. Philadelphia: University of Pennsylvania Press.
1977b The Myth of the Idle Peasant: Employment in a Subsistence Economy. *In* Peasant Livelihood: Studies in Economic Anthropology and Cultural Ecology. Rhoda Halperin and James Dow, eds. Pp. 60–78. New York: St. Martin's Press.
1993 Indigenous Knowledge of Biological Resources and Intellectual Property Rights: The Role of Anthropology. American Anthropologist 95(3):653–86.

Buechler, Hans C.
1969 Land Reform and Social Revolution in the Northern Altiplano and Yungas of Bolivia. *In* Land Reform and Social Revolution in Bolivia. Dwight B. Heath, Charles J. Erasmus, and Hans C. Buechler, eds. Pp. 169–240. New York: Frederick A. Praeger Publishers.

Burke, Melvin
1971 Land Reform in the Lake Titicaca Region. *In* Beyond the Revolution: Bolivia since 1952. James M. Malloy and Richard S. Thorn, eds. Pp. 301–340. Pittsburgh, PA: University of Pittsburgh Press.

Calderón Jemio, Raúl Javier
1991   In Defense of Dignity: The Struggles of the Aymara Peoples in the Bolivian Altiplano, 1830–1860. Ph.D. diss., University of Connecticut.

Campbell, Leon G.
1987   Ideology and Factionalism during the Great Rebellion, 1780–1782. *In* Resistance, Rebellion, and Consciousness in the Andean Peasant World, 18th to 20th Centuries. Steve J. Stern, ed. Pp. 110–139. Madison, WI: University of Wisconsin Press.

Carney, Heath, Michael W. Binford, and Alan L. Kolata
1996   Nutrient Fluxes and Retention in Andean Raised-Field Agriculture: Implications for Long-Term Sustainability. *In* Tiwanaku and Its Hinterland. Alan L. Kolata, ed. Pp. 169–179. Washington D.C.: Smithsonian Institution Press.

Carney, H. J., M. W. Binford, A. L. Kolata, R. Martin, and C. Goldman
1993   Nutrient and Sediment Retention in Andean Raised-Field Agriculture. Nature 364:131–33.

Carrier, James G.
1992a   Introduction. *In* History and Tradition in Melanesian Anthropology. J. Carrier, ed. Pp. 1–37. Berkeley, CA: University of California Press.
1992b   History and Tradition in Melanesian Anthropology. Berkeley, CA: University of California Press.
1992c   Occidentalism: The World Turned Upside-Down. American Ethnologist 19:195–212

Carter, William E.
1964   Aymara Communities and the Bolivian Agrarian Reform. Gainesville, FL: University of Florida Monographs No. 24.
1971   Revolution and the Agrarian Sector. *In* Beyond the Revolution: Bolivia since 1952. James M. Malloy and Richard S. Thorn, eds. Pp. 233–268. Pittsburgh, PA: University of Pittsburgh Press.

Chapin, Mac
1988   The Seduction of Models: Chinampa Agriculture in Mexico. Grassroots Development. 12(1):8–17.

Chibnik, Michael, and Wil de Jong
1989   Agricultural Labor Organization in Ribereño Communities of the Peruvian Amazon. Ethnology 28(1):75–95.

Clifford, James
1983   On Ethnographic Authority. Representations 1(2):118–46.

Clifford, James and George Marcus, eds.
1986   Writing Culture: The Poetics and Politics of Ethnography. Berkeley,
        CA: University of California Press.

Coe, Michael E.
1964   The *Chinampas* of Mexico. Scientific American 211:90–98.

Comaroff, John L.
1987   Of Totemism and Ethnicity: Consciousness, Practice and the Signs of
        Inequality. Ethnos 52(3–4):301–323.

Cook, Noble David
1981   Demographic Collapse: Indian Peru, 1520–1620. New York:
        Cambridge University Press.

Crespo V., Fernando
1993   Bolivia: Anuario Estadístico del Sector Rural, 1993. Centro de
        Información para el Desarrollo.

Dandler, Jorge and Juan Torrico A.
1987   From the National Indigenous Congress to the Ayopaya Rebellion:
        Bolivia, 1945–1947. *In* Resistance, Rebellion, and Consciousness in the
        Andean Peasant World, 18th to 20th Centuries. Steve J. Stern, ed. Pp.
        334–378. Madison, WI: University of Wisconsin Press.

Darch, J. P., ed.
1983   Drained Field Agriculture in Central and South America. B.A.R.
        International Series 189. Oxford, England.

Denevan, William M., Kent Mathewson, and Gregory Knapp, eds.
1987   Pre-Hispanic Agricultural Fields in the Andean Region. B.A.R.
        International Series 359 (ii). Oxford, England.

Denevan, William, and B. L. Turner II
1974   Forms, Functions, and Associations of Raised Fields in the Old World
        Tropics. Journal of Tropical Geography 39:24–33.

DeWalt, Billie R.
1994   Using Indigenous Knowledge to Improve Agriculture and Natural
        Resource Management. Human Organization 53(2):123–131.
1999   Combining Indigenous and Scientific Knowledge to Improve
        Agriculture and Natural Resource Management in Latin America. *In*
        Traditional and Modern Natural Resource Management in Latin
        America. F. J. Pichón, J. E. Uquillas, and J. Frechione, eds. Pittsburgh,
        PA: University of Pittsburgh Press.

Dirks, N., G. Eley, and S. Ortner
1994   Introduction. *In* Culture-Power-History: A Reader in Contemporary
        Social Theory. N. Dirks, G. Eley, and S. Ortner, eds. Princeton, NJ:
        Princeton University Press.

Dunkerley, James
1984   Rebellion in the Veins: Political Struggle in Bolivia, 1952–82. London:
        Verso Editions.

Erasmus, Charles
1956   Culture, Structure and Process: The Occurrence and Disappearance of
        Reciprocal Farm Labor. Southwestern Journal of Anthropology
        12:444–469.

Erickson, Clark L.
1985   Applications of Prehistoric Andean Technology: Experiments in Raised
        Field Agriculture, Huatta, Lake Titicaca: 1981–82. *In* Prehistoric
        Intensive Agriculture in the Tropics. Ian Farrington, ed. B.A.R.
        International Series 232, Part i.
1987   The Dating of Raised-Field Agriculture in the Lake Titicaca Basin,
        Peru. *In* Pre-Hispanic Agricultural Fields in the Andean Region.
        William Denevan, Kent Mathewson, and Gregory Knapp, eds. B.A.R.
        International Series 232, Part i.
1988a   An Archaeological Investigation of Raised Field Agriculture in the Lake
        Titicaca Basin of Peru. Ph.D. diss., University of Illinois at Urbana-
        Champaign
1988b   Raised Field Agriculture in the Lake Titicaca Basin: Putting Ancient
        Agriculture Back to Work. Expedition 30(3):8–16.
1992a   Applied Archaeology and Rural Development: Archaeology's Potential
        Contribution to the Future. Journal of the Steward Anthropological
        Society 20(1&2):1–16.
1992b   Prehistoric Landscape Management in the Andean Highlands: Raised
        Field Agriculture and Its Environmental Impact. Population and
        Environment 13(4):285–300.
1993   The Social Organization of Prehispanic Raised Field Agriculture in the
        Lake Titicaca Basin. *In* Research in Economic Anthropology:
        Economic Aspects of Water Management in the Pre-Hispanic New
        World. Vernon L Scarborough and Barry L. Isaac, eds. Supplement 7,
        Pp. 369–426. Greenwich, Connecticut: JAI Press.
1999   Neo-Environmental Determinism and Agrarian "Collapse" in Andean
        Prehistory. Antiquity 73:634–42.

Erickson, Clark L. and Daniel A. Brinkmeier
1991	Raised Field Rehabilitation Projects in the Northern Lake Titicaca Basin. A Report to the Inter-American Foundation. Unpublished Ms.

Erickson, Clark L. and Kay Candler
1989	Raised Fields and Sustainable Agriculture in the Lake Titicaca Basin of Peru. *In* Fragile Lands of Latin America: Strategies for Sustainable Development. John O. Bowder, ed. Pp. 230–48. Boulder, CO: Westview Press.

Escobar, Arturo
1995	Encountering Development: The Making and Unmaking of the Third World. Princeton, NJ: Princeton University Press.

Espinoza Soriano, W.
1980	Los Fundamentos Linguisticos de la Ethnohistoria Andina y Comentarios en Torno al Anonimo de Charcas de 1604. Revista Española de Antropologia Americana 10:149–69.

Farrington, I. S., ed.
1985	Prehistoric Intensive Agriculture in the Tropics. B.A.R. International Series 232, Parts i and ii.

Ferguson, James
1994	The Anti-Politics Machine: "Development," Depoliticization, and Bureaucratic Power in Lesotho. Minneapolis, MN: University of Minnesota Press.

Foucault, Michel
1972	The Archaeology of Knowledge. Translated by A. M. Sheridan Smith. New York: Pantheon Books.

Friedman, Jonathan
1992	The Past in the Future: History and the Politics of Identity. American Anthropologist 94(4):837–59.

Fundación Wiñaymarka
1995	Informe Final Parte I: Sukakollus Gestiones Agricolas 1987–1994. Report submitted to the Inter-American Foundation. July 1995.

Garaycochea Z., Ignacio
1987	Agricultural Experiments in Raised Fields in the Lake Titicaca Basin, Peru: Preliminary Considerations. *In* Pre-Hispanic Agricultural Fields in the Andean Region. William M. Denevan, Kent Mathewson, and Gregory Knapp, eds. Pp. 358–398. B.A.R. International Series 359, Part ii.

Gardner, Katy
1997   Mixed Messages: Contested 'Development' and the 'Plantation
        Rehabilitation Project.' *In* Discourses of Development: Anthropological
        Perspectives. R. D. Grillo and R. L. Stirrat, eds. Pp. 133–156. New
        York: Berg.

Gathercole, Peter and David Lowenthal, eds.
1990   The Politics of the Past. Boston: Unwin Hyman.

Goldstein, Paul S.
1989   The Tiwanaku Occupation of Moquegua. *In* Ecology, Settlement and
        History in the Osmore Drainage, Peru. Don S. Rice, Charles Stanish,
        and Phillip R. Scarr, eds. Pp. 219–256. B.A.R.: Oxford.

Godoy, Ricardo and Mario De Franco
1992   High Inflation and Bolivian Agriculture. Journal of Latin American
        Studies 24(3):617–637.

Gold, Mary V.
1994   Sustainable Agriculture: Definitions and Terms. Alternative Farming
        Systems Information Center Special Reference Briefs, SRB 94–05.
        Beltsville, MD: National Agricultural Library.

Graffam, Gray
1990   Raised Fields without Bureaucracy: An Archaeological Examination of
        Intensive Wetland Cultivation in the Pampa Koani Zone, Lake Titicaca,
        Bolivia. Ph.D. diss., University of Toronto.
1992   Beyond State Collapse: Rural History, Raised Fields, and Pastoralism in
        the South Andes. American Anthropologist 94(4):882–904.

Grieshaber, Erwin P.
1980   Survival of Indian Communities in 19th Century Bolivia: A Regional
        Comparison. Journal of Latin American Studies 12(2):223–269.

Guillet, David
1980   Reciprocal Labor and Peripheral Capitalism in the Central Andes.
        Ethnology 19(2):151–167.

Gupta, Akhil
1998   Postcolonial Developments: Agriculture in the Making of Modern
        India. Durham, NC: Duke University Press.

Hacking, Ian
1999   The Social Construction of What? Cambridge, MA: Harvard University
        Press.

Hale, Charles R.
1994   Between Che Guevara and the Pachamama: Mestizos, Indians and Identity Politics in the Anti-Quincentenary Campaign. Critique of Anthropology 14(1):9–39.

Hamilton, Kirk
1997   Accounting for Sustainability. *In* Approaches to Sustainable Development. Richard M. Auty and Katrina Brown, eds. Pp. 21–30. Washington, DC: Pinter.

Hamilton, Sarah
1998   The Two-Headed Household: Gender and Rural Development in the Ecuadorean Andes. Pittsburgh, PA: University of Pittsburgh Press.

Handler, Richard and Jocelyn Linnekin
1984   Tradition, Genuine or Spurious. Journal of American Folklore 97(385):273–90.

Hanson, Allan
1989   The Making of the Maori: Cultural Invention and its Logic. American Anthropologist 91:890–902.

Hardman, M.
1985   Aymara and Quechua: Languages in Contact. *In* South American Indian Languages. H. Manelis Klein and L. Stark, eds. Pp. 617–43 Austin, TX: University of Texas Press

Harris, Olivia
1995   Ethnic Identity and Market Relations: Indians and Mestizos in the Andes. *In* Ethnicity, Markets, and Migration in the Andes: At the Crossroads of History and Anthropology. Brooke Larson and Olivia Harris, eds., with Enrique Tandeter. Pp. 351–390. Durham, NC: Duke University Press.

Harwood, Richard R.
1990   A History of Sustainable Agriculture. *In* Sustainable Agricultural Systems. Clive A. Edwards et al. eds. Pp. 3–19. Delray Beach, FL: St. Lucie Press.

Healy, Kevin
1991   Political Ascent of Bolivia's Peasant Coca Leaf Producers. Journal of Interamerican Studies and World Affairs 33(spring):87–121.

Heath, Dwight B., Charles J. Erasmus, and Hans C. Buechler, eds.
1969   Land Reform and Social Revolution in Bolivia. New York: Frederick A. Praeger Publishers.

Herrera, Jesús
1980   Bolivia: Migraciones Internas Recientes segun el Censo Nacional de 1976. La Paz: INE

Hirst, K. Kris
1998a   Archaeology Interview: A lesson in Applied Archaeology (Interview with Clark Erickson). About.com Website. 10/20/00. <http://www. archaeology.about.com/science/archaeology/library/weekly/aa042698. htm?terms=Erickson&COB=home
1998b   Archaeology Interview: A lesson in Applied Archaeology. Part II: Recreating Raised Field Agriculture (Interview with Clark Erickson). About.com Website. 10/20/00. <http://www.archaeology.about.com /science/archaeology/library/weekly/aa050398.htm?terms=Erickson& COB=home
1998c   Archaeology Interview: A lesson in Applied Archaeology. Part III: Implications of the Research (Interview with Clark Erickson). About.com Website. 10/20/00. <http://www.archaeology.about.com/ science/archaeology/library/weekly/aa050398.htm?terms=Erickson&C OB=home

Hobsbawm, Eric
1983   Introduction: Inventing Traditions. *In* The Invention of Tradition. E. Hobsbawm and T. Ranger, eds. Pp. 1–14. New York: Cambridge University Press.

Hobsbawm, Eric and Terence Ranger, eds.
1983   The Invention of Tradition. New York: Cambridge University Press.

Humphries, S.
1993   Intensification of Traditional Agriculture Among Yucatec Maya Farmers: Facing up to the Dilemma of Livelihood Sustainablity. Human Ecology 21(1):87–102.

Hunn, Eugene S.
1999   The Value of Subsistence for the Future of the World. *In* Ethnoecology: Situated Knowledge/Located Lives. Virginia D. Nazarea, ed. Pp. 23–36. Tucson, AZ: University if Arizona Press.

INE (Instituto Nacional de Estadística)
1997   Bolivia: Proyecciones de Poblacion por Departamentos, segun Area Urban-Rural, Sexo y Grupos de Edad 1990–2010. LA Paz: INE.

Inter-American Foundation
1991   Amendment No. 3 to the Grant Agreement between the Parroquia Tiwanaku and the Inter-American Foundation. Grant No. BO-374–A3.

n.d.   Inter-American Foundation. 10/19/00. <http://www.iaf.gov/bouch/
          bouch.htm>.

n.d.   Inter-American Foundation. 10/19/00. <http://www.iaf.gov/loc
          dev-e.htm>.

International Union for Conservation of Nature and Natural Resources (IUCN)
1980   World Conservation Strategy: Living Resource Conservation for
          Sustainable Development. Published in Cooperation with the United
          Nations Environment Programme and the World Wildlife Fund.

Jackson, Robert H.
1989   The Decline of the Hacienda in Cochabamba, Bolivia: The Case of the
          Sacaba Valley, 1870–1929. Hispanic American Historical Review
          69(2):259–281.

Jacobsen, Nils
1993   Mirages of Transition: The Peruvian Altiplano, 1780–1930. Berkeley,
          CA: University of California Press.

Jameson, Kenneth P.
1989   Austerity Programs Under Conditions of Political Instability and
          Economic Depression: The Case of Bolivia. *In* Paying the Costs of
          Austerity in Latin America. Howard Handelman and Werner Baer, eds.
          Pp. 81–103. Boulder, CO: Westview Press.

Janusek, John Wayne
1994   State and Local Power in a Prehispanic Andean Polity: Changing
          Patterns of Urban Residence in Tiwanaku and Lukurmata, Bolivia.
          Ph.D. diss., The University of Chicago.

Jolly, Margaret
1992   Specters of Inauthenticity. The Contemporary Pacific 4:49–72.

Jones, James
1990   A Native Movement and March in Eastern Bolivia: Rationale and
          Response. Development Anthropology Network 8(2):1–8.

Kearney, Michael
1995   Introduction (Special Issue on "Ethnicity and Class in Latin America").
          Latin American Perspectives 23(2):5–16.
1996   Reconceptualizing the Peasantry: Anthropology in Global Perspective.
          Boulder, CO: Westview Press.

Kehoe, Alice B.
1996   Participant Observation with the Lakaya Centro de Madres. *In*
          Tiwanaku and Its Hinterland. Alan L. Kolata, ed. Pp. 231–40.
          Washington D.C.: Smithsonian Institution Press.

Kellogg, Susan
1991   Histories for Anthropology: Ten Years of Historical Research and Writing by Anthropologists, 1980–1990. Social Science History 15(4):417–455.

Kempton, Willett, James S. Boster, and Jennifer A. Hartley
1995   Environmental Values in American Culture. Cambridge, MA: MIT Press.

Klein, Herbert S.
1982   Peasant Response to the Market and the Land Question in 18th and 19th Century Bolivia. Nova Americana 5:3–127.
1992 (1982)   Bolivia: The Evolution of a Multi-Ethnic Society (second edition). New York: Oxford University Press.
1993   Haciendas and Ayllus: Rural Society in the Bolivian Andes in the 18th and 19th Centuries. Stanford, CA: Stanford University Press.

Kolata, Alan L.
1986   The Agricultural Foundations of the Tiwanaku State: A View from the Heartland. American Antiquity 51(4):748–762.
1991   The Technology and Organization of Agricultural Production in the Tiwanaku State. Latin American Antiquity 2(2):99–125.
1993   The Tiwanaku: Portrait of an Andean Civilization. Cambridge, MA: Blackwell Publishers.
1992   Economy, Ideology, and Imperialism in the South-Central Andes. In Ideology and Pre-Columbian Civilizations. A. Demerest and G. Conrad, eds. Pp. 65–85. Santa Fe, NM: School of American Research Press.
1996a   Proyecto Wila Jawira: An Introduction to the History, Problems, and Strategies of Research. In Tiwanaku and Its Hinterland. Alan L. Kolata, ed. Pp. 1–22. Washington D.C.: Smithsonian Institution Press.
1996b   Tiwanaku and Its Hinterland. Alan L. Kolata, ed. Washington D.C.: Smithsonian Institution Press.
1996c   Valley of the Spirits: A Journey into the Lost Realm of the Aymara. New York: John Wiley & Sons, Inc.

Kolata, Alan L., and Charles R. Ortloff
1989   Thermal Analysis of Tiwanaku Raised Field Systems in the Lake Titicaca Basin of Bolivia. Journal of Archaeological Science 16:233–263.
1996   Tiwanaku Raised-Field Agriculture in the Lake Titicaca Basin of Bolivia. In Tiwanaku and Its Hinterland. Alan L. Kolata, ed. Pp.109–151. Washington, D.C.: Smithsonian Institution Press.

Kolata, Alan L., Oswaldo Rivera, Juan Carlos Ramírez, and Evelyn Gemio
1996   Rehabilitating Raised-Field Agriculture in the Southern Lake Titicaca Basin of Bolivia: Theory, Practice, and Results. *In* Tiwanaku and Its Hinterland. Alan L. Kolata, ed. Pp. 203–230. Washington, D.C.: Smithsonian Institution Press.

Kozloff, Robin Rose
1994   Community Factors in Agricultural Change: The Introduction of Raised Fields in Highland Bolivia. Unpublished M.S. Thesis.

Krech, Shepard, III.
1999   The Ecological Indian: Myth and History. New York: W. W. Norton & Company.

Lagos, Maria L.
1994   Autonomy and Power: The Dynamics of Class and Culture in Rural Bolivia. Philadelphia: University of Pennsylvania Press.

Langer, Erick D.
1989   Economic Change and Rural Resistance in Southern Bolivia, 1880–1930. Stanford, CA: Stanford University Press.

Larson, Brooke
1998 (1988)   Cochabamba, 1550–1900: Colonialism and Agrarian Transformation in Bolivia. Durham, NC: Duke University Press.

Lindstrom, Lamont and Geoffrey M. White, eds.
1994   Culture, Kastom, Tradition: Developing Cultural Policy in Melanesia. Suva, Fiji: Institute of Pacific Studies, University of the South Pacific.

Lowenthal, David
1990   Conclusion: Archaeologists and Others. *In* The Politics of the Past. P. Gathercole and D. Lowenthal, eds. Pp.302–14. Boston: Unwin Hyman.

Lovelock, James E.
1987   Gaia: A New Look at Life on Earth. New York: Oxford University Press.

Mallon, Florencia E.
1983   The Defense of Dignity in Peru's Central Highlands: Peasant Struggle and Capitalist Transition, 1860–1940. Princeton, NJ: Princeton University Press.

Malloy, James M. and Richard S. Thorn, eds.
1971   Beyond the Revolution: Bolivia Since 1952. Pittsburgh, PA: University of Pittsburgh Press.

198   Bibliography

Mamani Condori, Carlos
1989   History and Prehistory in Bolivia: What About the Indians? *In* Conflict
        in the Archaeology of Living Traditions. Robert Layton, ed. Pp. 46–59.
        Boston, MA: Unwin Hyman.

Marcus, George E. and Michael M. J. Fischer
1986   Anthropology as Cultural Critique: An Experimental Moment in the
        Human Sciences. Chicago, IL: The University of Chicago Press.

Mathewson, Kent
1987   Estimating Labor Inputs for the Guayas Raised Fields: Initial
        Considerations. *In* Pre-Hispanic Agricultural Fields in the Andean
        Region. W. Denevan, K. Mathewson, and G. Knapp, eds. Pp.321–36.
        BAR International Series 359 (ii).

McCarthy, E. Doyle
1996   Knowledge as Culture: The New Sociology of Knowledge. New York:
        Routledge.

McCormick, John
1989   Reclaiming Paradise: The Global Environmental Movement.
        Bloomington, IN: Indian University Press.

Meerman, Jacob P.
1997   Reforming Agriculture: The World Bank Goes to Market. World Bank
        Operations Evaluation Study. Washington, DC: The World Bank.

Meja, Volker, and Nico Stehr
1990   On the Socio9logy of Knowledge Dispute. *In* Knowledge and Politics:
        The Sociology of Knowledge Dispute. Volker Meja and Nico Stehr, eds.
        Pp. 3–13. New York: Routledge.

Mendelberg, Uri
1985   The Impact of the Bolivian Agrarian Reform on Class Formation. Latin
        American Perspectives 12(3):45–58.

Minesterio de Asuntos Campesinos y Agropecuarios
1991   Importaciones Agropecuarias de Bolivia 1981–1989. La Paz.

Mitchell, William P.
1991   Some are More Equal than Others: Labor Supply, Reciprocity, and
        Redistribution in the Andes. Research in Economic Anthropology
        13:191–219.

Morales, Juan Antonio
1991   Structural Adjustment and Peasant Agriculture in Bolivia. Food Policy
        16:58–66.

Morales, Juan Antonio and Jeffrey D. Sachs
1989   Bolivia's Economic Crisis. *In* Developing Country Debt and the World Economy. Jeffrey Sachs, ed. Pp. 57–79. Chicago: University of Chicago Press.

Moran, Katy
1999   Toward Compensation: Returning Benefits from Ethnobotanical Drug Discovery to Native Peoples. *In* Ethnoecology: Situated Knowledge/Located Lives. Virginia D. Nazarea, ed. Pp. 249–62. Tucson, AZ: University if Arizona Press.

Moseley, Michael E.
1992   The Incas and their Ancestors. New York: Thames and Hudson.

Murra, John V.
1968   An Aymara Kingdom in 1567. Ethnohistory 15:115–51.
1984   Andean Societies before 1532. *In* The Cambridge History of Latin America, Volume one: Colonial Latin American. Leslie Bethell, ed. Pp. 59–90. New York: Cambridge University Press.
1986   The Expansion of the Inka State: Armies, War, and Rebellions. *In* Anthropological History of Andean Polities. John V. Murra, Nathan Watchel, and Jacques Revel, eds. Pp. 49–58. New York: Cambridge University Press.
1995   Did Tribute and Markets Prevail in the Andes before the European Invasion? *In* Ethnicity, Markets, and Migration in the Andes: At the Crossroads of History and Anthropology. Brooke Larson and Olivia Harris with Enrique Tandeter, eds. Pp. 57–72. Durham, NC: Duke University Press.

Nash, June
1992   Interpreting Social Movements: Bolivian Resistance to the Economic Conditions Imposed by the International Monetary Fund. American Ethnologist 19(2):275–293.

Nazarea, Virginia D.
1999   Introduction: A View from a Point: Ethnoecology as Situated Knowledge. In Ethnoecology: Situated Knowledge/Located Lives. Virginia D. Nazarea, ed. Pp.3–20. Tucson, AZ: University of Arizona Press.

Netting, Robert McC.
1993   Smallholders, Householders: Farm Families and the Ecology of Intensive, Sustainable Agriculture. Stanford, California: Stanford University Press.

Nordholt, H. S.
1994   The Making of Traditional Bali: Colonial Ethnography and Bureaucratic Reproduction. History and Anthropology 8(1):89–127.

Nugent, David
1997   Modernity at the Edge of Empire: State, Individual, and Nation in the Northern Peruvian Andes, 1885–1935. Stanford, CA: Stanford University Press.

Nygren, Amja
1999   Local Knowledge in the Environment-Development Discourse: From Dichotomies to Situated Knowledges. Critique of Anthropology 19(3):267–288.

Organization for Economic Co-operation and Development (OECD)
1995   Sustainable Agriculture: Concepts, Issues and Policies in OECD Countries. Paris: OECD.

Ortloff, Charles R.
1996   Engineering Aspects of Tiwanaku Groundwater-Controlled Agriculture. *In* Tiwanaku and Its Hinterland. Alan L. Kolata, ed. Pp. 153–167. Washington, D.C.: Smithsonian Institution Press.

Ortloff, Charles R. and Alan L. Kolata
1989   Hydraulic Analysis of Tiwanaku Aqueduct Structures at Lukurmata and Pajchiri, Bolivia. Journal of Archaeological Science 16:513–535.
1993   Climate and Collapse: Agro-Ecological Perspectives on the Decline of the Tiwanaku State. Journal of Archaeological Science 20:195–221.

Ortner, Sherry
1984   Theory in Anthropology since the Sixties. Comparative Studies in Society and History 26(1):126–166.
1989   High Religion: A Cultural and Political History of Sherpa Buddhism. Princeton, NJ: Princeton University Press.
1991   Reading America: Preliminary Notes on Class and Culture. *In* Recapturing Anthropology: Working in the Present. Richard Fox, ed. Pp. 136–89. Santa Fe, NM: School of American Research Press.

Painter, Michael
1991   Re-creating Peasant Economy in Southern Peru. *In* Golden Ages, Dark Ages: Imagining the Past in Anthropology and History. J. O'Brien and William Roseberry, eds. Pp. 81–106. Berkeley: University of California Press.

Parsons, J. J. and W. A. Bowen
1967   Pre-Columbian Ridged Fields. Scientific American 217:92–100.

Patterson, Thomas C.
1999   The Political Economy of Archaeology in the United States. Annual Review of Anthropology 28:155–74.

Pezzey, J.
1989   Economic Analysis of Sustainable Growth and Sustainable Development. Environment Department Working Paper No 15. Washington, DC: The World Bank.

Platt, Tristan
1982   Estado Boliviano y Ayllu Andino: Tierra y Tributo en el Norte de Potosí. Lima: Instituto de Estudios Peruanos.
1987   The Andean Experience of Bolivian Liberalism, 1825–1900: Roots of Rebellion in 19th-Century Chayanta (Potosí). *In* Resistance, Rebellion, and Consciousness in the Andean Peasant World, 18th to 20th Centuries. Steve J. Stern, ed. Pp. 280–323. Madison, WI: University of Wisconsin Press.
1993   Simón Bolívar, the Sun of Justice and the Amerindian Virgin: Andean Conceptions of the *Patria* in 19th Century Potosí. Journal of Latin American Studies. 25:159–185.

Posey, Darrell A.
1999   Safeguarding Traditional Resource Rights of Indigenous Peoples. *In* Ethnoecology: Situated Knowledge/Located Lives. Virginia D. Nazarea, ed. Pp. 217–29. Tucson, AZ: University if Arizona Press.

Purcell, Trevor W.
1998   Indigenous Knowledge and Applied Anthropology: Questions of Definition and Direction. Human Organization 57(3):258–272.

Quiroz, Consuelo
1999   Local Knowledge Systems (LKS) in Latin America: Current Trends and Contributions Towards Sustainable Development. *In* Traditional and Modern Natural Resource Management in Latin America. F. J. Pichón, J. E. Uquillas, and J. Frechione, eds. Pittsburgh, PA: University of Pittsburgh Press.

Rabinow, Paul
1986   Representations are Social Facts: Modernity and Post-Modernity in Anthropology. *In* Writing Culture: The Poetics and Politics of Ethnography. James Clifford and George Marcus, eds. Pp. 234–261. Los Angeles, CA: University of California Press.

Rivera Cusicanqui, Silvia
1978   La Expanción del Latifundio en el Altiplano Boliviano: Elementos para

la Caracterización de una Oligarquía Regional. Avances, No. 2. La Paz.
1987   "Oppressed but not Defeated": Peasant Struggles among the Aymara
and Quechua in Bolivia, 1900–1980. Geneva: United Nations Research
Institute for Social Development.
1993   Anthropology and Society in the Andes: Themes and Issues. Critique of
Anthropology 13(1):77–96.

Robbins, Nicholas
1994   Revivalist Nativism in the Andean Highlands: The 1780–1781
Rebellion in Upper Peru. PhD. diss., Tulane University.

Rojas-Velarde, Luis
1996   *Informe de Monitoreo.* Monitor Report on *Parroquia de Tiwananku –
Fundación Wiñaymarka* submitted to the Inter-American Foundation.

Sahlins, Marshal
1965   On the Sociology of Primitive Exchange. *In* The Relevance of Models
for Social Anthropology. Michael Banton, ed. Pp. 139–236. Association
for Social Anthropologists, Monograph 1. New York: Praeger.

Sagarnaga M., J. Antonio
1991   Tiwanaku: The Map and Guide. La Paz: Editorial Quipus.

Said, Edward W.
1978   Orientalism. New York: Vintage Books.

Saignes, Thierry
1995   Indian Migration and Social Change in 17th Century Charcas. *In*
Ethnicity, Markets, and Migration in the Andes: At the Crossroads of
History and Anthropology. Brooke Larson and Olivia Harris with
Enrique Tandeter, eds. Pp. 167–195. Durham, NC: Duke University
Press.

Salomon, Frank
1982   Andean Ethnology in the 1970s: A Retrospective. Latin American
Research Review 17(2):75–128.

Sanabria, Harry
1993   The Coca Boom and Rural Social Change in Bolivia. Ann Arbor, MI:
University of Michigan Press.

Sánchez de Lozada, Diego
1996   Heat and Moisture Dynamics in Raised Fields of the Lake Titicaca
Region, Bolivia. Ph.D. diss., Cornell University.

San Martín Arzabe, Hugo
1991   El Palenquismo: Movimiento Social, Populismo, Informalidad Política. La Paz: Los Amigos del Libro.

Schmidt, Peter R.
1995   Using Archaeology to Remake History in Africa. *In* Making Alternative Histories: The Practice of Archaeology and History in Non-Western Settings. P. Schmidt and T. Patterson, eds. Pp. 119–47. Santa Fe, NM: School of American Research Press.

Schmidt, Peter R. and Thomas C. Patterson.
1995   Introduction: From Constructing to Making Alternative Histories. *In* Making Alternative Histories: The Practice of Archaeology and History in Non-Western Settings. P. Schmidt and T. Patterson, eds. Pp. 1–24. Santa Fe, NM: School of American Research Press.

Seddon, Mathew Thomas
1994   Excavations in the Raised Fields of the Rio Catari Sub-Basin, Bolivia. Unpublished M.A. Thesis. University of Chicago.

Sillitoe, Paul
1998   The Development of Indigenous Knowledge: A New Applied Anthropology. Current Anthropology 39(2):223–252.

Smith, C.T., W.M. Denevan, and P. Hamilton
1968   Ancient Ridged Fields in the Region of Lake Titicaca. Geographical Journal 134(3):353–367.

Smith, Kimbra
2000   'We Paint Just Like Our Ancestors': The Discursive Appropriation of Pacha in Cajamarca, Peru. Paper presented at the American Ethnological Society Annual Meeting, March 24–26, Tampa, FL.

Soleri, Daniela and Steven E. Smith
1999   Conserving Folk Crop Varieties: Different Agricultures, Different Goals. *In* Ethnoecology: Situated Knowledge/Located Lives. Virginia D. Nazarea, ed. Pp. 133–54. Tucson, AZ: University if Arizona Press.

Stanish, Charles
1992   Ancient Andean Political Economy. Austin, TX: University of Texas Press.
1994   The Hydraulic Hypothesis Revisited: Lake Titicaca Basin Raised Fields in Theoretical Perspective. Latin American Antiquity 5(4):312–32.

Stephenson, David J. Jr.
1999   A Practical Primer on Intellectual Property Rights in a Contemporary

Ethnological Context. *In* Ethnoecology: Situated Knowledge/Located Lives. Virginia D. Nazarea, ed. Pp. 230–48. Tucson, AZ: University if Arizona Press.

Stern, Steve J

1987a   The Age of Andean Insurrection, 1742–1782: A Reappraisal. *In* Resistance, Rebellion, and Consciousness in the Andean Peasant World, 18th to 20th Centuries. Steve J. Stern, ed. Pp. 34–93. Madison, WI: University of Wisconsin Press.

1987b   New Approaches to the Study of Peasant Rebellion and Consciousness: Implications of the Andean Experience. *In* Resistance, Rebellion, and Consciousness in the Andean Peasant World, 18th to 20th Centuries. Steve J. Stern, ed. Pp. 3–25. Madison, WI: University of Wisconsin Press.

1995   The Variety and Ambiguity of Native Andean Intervention in European Colonial Markets. *In* Ethnicity, Markets, and Migration in the Andes: At the Crossroads of History and Anthropology. Brooke Larson and Olivia Harris with Enrique Tandeter, eds. Pp. 73–100. Durham, NC: Duke University Press.

Swartley, Lynn

n.d.   Demography, Political Economy, and Agricultural Intensification: Shifting Patterns of Land Use in the Lake Titicaca Basin of Bolivia in the 19th and 20th Centuries. Unpublished Ms.

Szeminski, Jan

1987   Why Kill the Spaniard? *In* Resistance, Rebellion, and Consciousness in the Andean Peasant World, 18th to 20th Centuries. Steve J. Stern, ed. Pp. 166–192. Madison, WI: University of Wisconsin Press.

Tapia N., Mario E. and Mariano Banegas

1990   Human Adaptations to a High-Risk Environment: *Camellones* or *Waru Waru*—Traditional Agricultural Technology of the Peruvian Andes. Journal of Farming Systems Research-Extension 1(1):93–98.

Terero, A

1992   Acerca de la Familia Linguistica Uruquilla. Revista Andina. 19:171–91.

Thomas, Nicholas

1992   Substantivization and Anthropological Discourse: The Transformation of Practices into Institutions in Neotraditional Pacific Societies. *In* History and Tradition in Melanesian Anthropology. J. Carrier, ed. Pp. 64–85. Berkeley, CA: University of California Press.

Thomson, Sinclair
1996   Colonial Crisis, Community, and Andean Self-Rule: Aymara Politics in the Age of Insurgency (18th Century La Paz). Ph.D. diss., University of Wisconsin-Madison.

Thorn, Richard S.
1971   The Economic Transformation. *In* Beyond the Revolution: Bolivia since 1952. James M. Malloy and Richard S. Thorn, eds. Pp.157–216. Pittsburgh: University of Pittsburgh Press.

Tschopik, Harry, Jr.
1946   The Aymara. *In* Handbook of South American Indians, Volume 2. Julian Steward, ed. Pp. 501–73. Washington: Smithsonian Institution.

Trouillot, Michel-Rolph
1991   Anthropology and the Savage Slot: The Poetics and Politics of Otherness. *In* Recapturing Anthropology: Working in the Present. Richard G. Fox, ed. Pp. 17–44. Santa Fe, NM: School of American Research Press.

Turner, B. L. II
1974   Prehistoric Intensive Agriculture in the Mayan Lowlands. Science 185:118–124.

Turner, B.L. II, and William M. Denevan
1985   Prehistoric Manipulation of Wetlands in the Americas: A Raised Field Perspective. *In* Prehistoric Intensive Agriculture in the Tropics. I. S. Farrington, ed. B.A.R. International Series 232, part i.

Valle de Siles, María Eugenia
1990   Historía de la Rebelíon de Túpac Catari, 1781–1782. La Paz, Bolivia: Editorial Don Bosco.

Van Cott, Donna Lee
2000a   Bolivia: The Construction of a Multiethnic Democracy. *In* Latin American Politics and Development, 5th edition . Howard J. Wiarda and Harvey Kline, eds. Boulder, CO: Westview Press.
2000b   The Friendly Liquidation of the Past: The Politics of Diversity in Latin America. Pittsburgh, PA: University of Pittsburgh Press.

Wachtel, Nathan
1986   Men of the Water: The Uru Problem (16th and 17th centuries). *In* Anthropological History of Andean Polities. John V. Murra, Nathan Watchel, and Jacques Revel, eds. Pp. 283–310. New York: Cambridge University Press.

Wallerstein, Immanuel
1974   The Modern World-System I: Capitalist Agriculture and the Origins of the European World-Economy in the Sixteenth Century. New York: Academic Press.

Warren, D. Michael
1989a   Linking Scientific and Indigenous Agricultural Systems. *In* The Transformation of International Agricultural Research and Development. J. Lin Crompton, ed. Pp. 153–170. Boulder, CO: Lynne Rienner Publishers.
1989b   Utilizing Indigenous Healers in National Health Delivery Systems: The Ghanaian Experiment. *In* Making Our Research Useful: Case Studies in the Utilization of Anthropological Knowledge. John van Willigen, ed. Pp. 159–78. Boulder, CO: Westview Press.
1991   Using Indigenous Knowledge in Agricultural Development. World Bank Discussion Papers No. 127. Washington, DC: The World Bank.
1999   Indigenous Knowledge for Agricultural Development. *In* Traditional and Modern Natural Resource Management in Latin America. F. J. Pichón, J. E. Uquillas, and J. Frechione, eds. Pittsburgh, PA: University of Pittsburgh Press.

Wessel, Kelso L.
1966   Social-Economic Comparison of Eight Agricultural Communities in the Oriente and the Altiplano. Department of Agricultural Economics of Cornell University. Unpublished Ms.

Wolf, Eric R.
1955   Types of Latin American Peasantry: A Preliminary Discussion. American Anthropologist 58:1065–78.

World Bank
1997   Advancing Sustainable Development: The World Bank and Agenda 21. Environmentally Sustainable Development Studies and Monographs Series No. 19. Washington, DC: The World Bank.

Worsley, Peter
1997   Knowledges: Culture-Counterculture-Subculture. New York: The New Press.

Index

.

For Product Safety Concerns and Information please contact our EU
representative GPSR@taylorandfrancis.com
Taylor & Francis Verlag GmbH, Kaufingerstraße 24, 80331 München, Germany

www.ingramcontent.com/pod-product-compliance
Lightning Source LLC
Chambersburg PA
CBHW050434280326
41932CB00013BA/2108

9 7 8 1 1 3 8 9 7 3 3 1 2